The culture of technology needs moral guidance, the philosophy of technology needs new life. Michel Puech's book gives us both.

Albert Borgmann, *University of Montana, USA*

The question of how to live a good life—a pressing question for any
thoughtful person—has taken on a particular urgency as the pace of technological change increasingly configures the world we wake up to every day.
In this bold, lucid work, Michel Puech proposes approaching this question
by looking to a realm traditionally neglected by many philosophers as
worthy of serious attention: ordinary life itself. Deftly supporting his
analysis with his extensive knowledge of diverse philosophical traditions,
Puech brings the familiar world of everyday "micro-actions", such as
texting, driving, and making coffee, before our eyes in a fresh light,
showing how they can promote flourishing without promoting compla-
cency or preventing resistance to technology when appropriate. Written in
a spirit of intellectual joy, this is an important volume not only for ethicists
and philosophers of technology, but for all with an inquiring mind.

Diane P. Michelfelder, *Macalester College, USA*

The Ethics of Ordinary Technology

Technology is even more than our world, our form of life, our civilization. Technology interacts with the world to change it. Philosophers need to seriously address the fluidity of a smartphone interface, the efficiency of a Dyson vacuum cleaner, or the familiar noise of an antique vacuum cleaner. Beyond their phenomenological description, the emotional experience acquires moral significance and in some cases even supplies ethical resources for the self. If we leave this dimension of modern experience unaddressed, we may miss something of value in contemporary life.

Combining European humanism, Anglophone pragmatism, and Asian traditions, Michel Puech pleads for an "ethical turn" in the way we understand and address technological issues in modern-day society. Puech argues that the question of "power" is what needs to be reconsidered today. In doing so, he provides a three-tier distinction of power: power to modify the outer world (our first-intention method in any case: technology); power over other humans (our enduring obsession: politics and domination); power over oneself (ethics and wisdom).

Michel Puech is Associate Professor of Philosophy at Paris-Sorbonne University, France. His current work focuses on the notion of modern wisdom, combining philosophy of technology, applied ethics, and Asian schools of thought.

Routledge Studies in Science, Technology and Society

The Ethics of Ordinary Technology

Michel Puech

Routledge
Taylor & Francis Group

NEW YORK AND LONDON

First published 2016
by Routledge
711 Third Avenue, New York, NY 10017

and by Routledge
2 Park Square, Milton Park, Abingdon, Oxon OX14 4RN

First issued in paperback 2017

Routledge is an imprint of the Taylor & Francis Group, an informa business

Library of Congress Cataloging in Publication Data
Names: Puech, Michel.
Title: The ethics of ordinary technology / Michel Puech.
Description: New York : Routledge, 2016. | Series: Routledge studies in science, technology and society ; 33 | Includes bibliographical references and index.
Identifiers: LCCN 2015045441| ISBN 9781138659346 (hbk) | ISBN 9781315620282 (ebk)
Subjects: LCSH: Technology–Moral and ethical aspects. | Technology and civilization. | Technology–Philosophy.
Classification: LCC BJ59 .P84 2016 | DDC 174/.96–dc23
LC record available at http://lccn.loc.gov/2015045441

ISBN 13: 978-1-138-48654-6 (pbk)
ISBN 13: 978-1-138-65934-6 (hbk)

Typeset in Sabon
by Wearset Ltd, Boldon, Tyne and Wear

Contents

Preface

This book is not simply a program for a future ethics of technology. Its ambition is to execute this kind of program and thus to give one version of a philosophy and ethics of contemporary technology. It aims at a pragmatic ethics, a form of applied ethics that really impacts personal and collective actions in the real world. The arguments in this book are then to be endorsed by the individual, in one's own name. They neither intend to influence any institution nor to found yet another academic branch. This method requires engagement and the risk of holding a stance, even if mitigated by a sincere effort to provide cogent arguments for every choice, as far as possible. The reader may also benefit from this research in an ethically admirable way: by reaching different conclusions.

The order of the chapters in this book is the most natural one for a philosophical reader. Another reading order is possible: starting from Chapter 6 and backward to Chapter 1 – starting with pragmatic wisdom practices, then exploring their context, and ending with the methods.

The most important sections for the structure and argumentation of this book are:

1.2 and 1.3 about methods
2.3 on coevolution (humans, nature, technology) and its potentials
3.2 on the ethical importance of the ordinary
4.2 on the ethical focus on the self
5.2 on the theory of microactions
6.1 on the paradigm of wisdom in a technoethical context.

The most concrete and applied sections are:

2.1 on the natural technosphere
3.1 on the ordinary technosphere
3.4 on ordinary care and attachment
4.4 on the good life in technology
5.3 on inclusive ordinary technology
6.3 and 6.4 on wisdom practical virtues and skills in the technosphere.

Acknowledgments

For their help in writing and publishing this book, I am grateful to Albert Borgmann, Adélaïde de Lastic, Christelle Didier, Christine Genet, Pietro Majno, Diane Michelfelder, Carl Mitcham, Natalja Mortensen, Roberto Toledo.

1 An Introduction to Technoethics

1.1 Why Ethics About Technology and How?

Modernity/Technology

The manifest characteristic of our contemporary world is technology. This statement engages a scale of value and significance that can eventually be reduced to pure and simple ideological prejudice by some cultural relativist, but its *prima facie* obviousness stands firm as a starting point for our inquiry. Modernity is perceived as technology. This fact is a question, the famous "Question concerning technology" elaborated by Martin Heidegger (1954). Philosophers of technology have never ceased, in the last decades of the twentieth century, to pose again this question. Langdon Winner (1986: Chapter 1) identifies technology as our typical "form of life," in the sense of Wittgenstein, that is to say a systemic cultural and existential pattern. Albert Borgmann's book titled *Technology and the character of contemporary life* (Borgmann 1984) is a milestone in philosophy. On this basis, the fact that technology is the main characteristic of contemporary life and world has the status of a definition in philosophy of technology.

The most important sociological approaches to modernity stress the importance of technology in terms that are cognate to "ordinary technoethics":

> Modernity is a post-traditional order, in which the question, "How shall I live?" has to be answered in day-to-day decisions about how to behave, what to wear, and what to eat – and many other things – as well as interpreted within the temporal unfolding of self-identity.
>
> (Giddens 1991: 14)

More precisely, the ethical content that has long been implicit in philosophy of technology is now emerging, either in a constructive approach (Brey et al. 2012: 1–3) or in a critical approach centered on the crisis of modernity as an ideal (Toulmin 1990). Technological superiority is still taken as an absolute civilizational superiority, in the West at least, but through value assumptions that can hardly claim universality (Mary Tiles in Hershock et al. 2003: 493).

Assessing technology is a methodological challenge. Every contemporary questioning is informed by a technological preconception (Böhme 2012), in a sense reminiscent of Heidegger's basic argument: technology is already present in every questioning frame, so there is a real obstacle to frame technology itself as a problem. The most important consequence of this methodological issue is, in Borgmann's terms, "the immunity of technology to traditional moral analysis" (Borgmann 1984: 169).

Technology is even more than our world, our form of life, our civilization. We need to track down a more intimate presence of technology in our world and our life. Don Ihde's work has highlighted this existential proximity of technology. "Technology makes the 'texture' of our world and of our existence" (Ihde 1979: 63); "our existence is 'technically textured'" (Ihde 1990: 1). This technological texture is on the objective side what I will call the *technosphere:* our actual environment, with the pervasive presence of ambient and background technology (Ihde 1979: Chapter 5). But there is a second and more intimate layer, which I will call the *proximal technosphere*. It is on the subjective side, or more precisely, at the interface between subject and world. Recent philosophy of technology converges toward the analysis of a *technologically mediated self*. I will argue that this existential significance of technology is the new frame in which its ethical significance can be analyzed.

There is between science and technology an ontological difference with heavy ethical consequences. Science is discourse; technology is action. Science builds a representation of the world; technology interacts with the world to change it. The values of science are representational values. They are primarily "truth" values or, better, "validation" procedures. The values of technology are necessarily agency values, that is to say, ethical values. The specific values system of technology originates in this rupture with the central epistemic value of the West: truth (Kitcher 2001: 103, 166; Pitt 1995).

The second point to clarify is the precise meaning of non-neutrality in technoethics. I will insist on the oracular Kranzberg's (First) Law: "Technology is neither good nor bad; nor is it neutral" (Kranzberg 1986). Non-neutrality in technology does not mean that technology has by essence a definite value, be it good or bad. This simplistic valuation opens the way to dualistic ideologies: technophilia and technophobia. On the contrary, non-neutrality means that technology always imports and sometimes imposes a values system that makes a real difference, but that never supplies predetermined moral categories such as good or bad. To have a knife is neither good nor bad, it can be used indifferently as a powerful tool or as a weapon, but to have a knife is not neutral, in any circumstance. It modifies the global situation. It creates a context and a unique moral context. How can we capture this situation and context?

A typical uneasiness looms over every attempt to understand the moral in a given technological context. Yes, the suspect was carrying a knife for innocent picnic uses, but she never excluded the possibility (how could she?) of using it as a dissuading weapon in case of a bad encounter during this solitary hiking expedition. Yes, that particular kind of knife is more than you need to cut your bread, but it is the appropriate knife to be clipped to a belt and possibly used as a tool in a wild nature environment. How can we morally frame human intentions and artifacts affordances (potential uses) together? The inadequate cultural integration of technology and in particular the uneasy ethical assessment of technology is for Hottois a consequence of maintaining *technē* and *logos* as two separate orders of reality (Hottois 1984). The symbolic and value-laden agency of humans *(logos)* is kept apart from technical artifacts, which maintains technology in an illusory neutrality or invisibility and obstructs the vision of a global and hybrid framework in which *logos* and *technē* are jointly deployed in human projects.

In technology itself, there is a values system to elucidate and to assess. This task can be carried out and it has been carried out with different outcomes, on three levels:

1   *Political critique.* The power of a military-industrial complex and its hidden agenda have been exposed by Jacques Ellul (1954) and this has been the basis of a political critique of industrial technology from the 1970s on. The crude version of this approach is more than often a conspiracy theory where technology is instrumentalized by the "bad guys," but more realistic versions can be found in Herbert Marcuse (1964) and in philosophers of technology like Winner (1977) or Feenberg (1999, 2002).
2   *Ethical assessment.* Engineering ethics and some branches of applied ethics have produced several models of values analysis in technology. These models will be a substantial source for my argument.
3   *Existential awareness.* To address the relevant level of intimacy that we maintain with our contemporary artifacts, the methods of ethical assessment inspired by engineering ethics and standard applied ethics can be evolved into a form of existential analysis with a stronger interpretative ambition and a resolute focus on the self. The objective of this approach is to delineate an existential non-neutrality of technology, more important than ethical non-neutrality (Kranzberg's Law) and key to some of the yet poorly addressed issues of modernity. Borgmann's work pioneered this way (Borgmann 1984) in philosophy of technology, and Michel Foucault provided indispensable guidance to carry on this task in technoethics.

From Philosophy of Technology to Technoethics

Applied ethics embraces a variety of domains, corresponding to the various activities of human agents: business, medicine, engineering, law, war, education, interactions with the natural environment, the digital environment, and so on (Frey and Wellman 2003). Bioethics constitutes a large and relatively coherent division of applied ethics, at least under a broad definition of bioethics that includes medical and environmental ethics plus the ethics of research and development in biology and biotechnology. Technoethics may reasonably try to become an equivalent division of applied ethics, under a broad definition that includes any interaction with artifacts (human-made objects, see Baker 2004; Hilpinen 2011; Olsen et al. 2009: Chapter 28). Bioethics and technoethics both require efforts to assess the philosophical consequences of their two foundational concepts, life and artifact. To put it mildly, artifacts have received less philosophical attention.

The introduction to philosophy of technology by Tiles and Oberdiek (1995) remains a sound approach to the field in its entirety. It typically includes the consideration of values. Technology is not just applied science. From this founding observation specific problems arise, which have been neglected, say Tiles and Oberdiek. To address these problems requires a break from the pro- or anti-technology valuation systems that rule when the question of technology is only an excuse for a critique or for a promotion of preexisting social values. But in the much needed bigger picture, where is technology to be situated in relation to the major ontological and ethical distinctions *facts/values* and *human/nature* (Tiles and Oberdiek 1995: 29–30)?

Mitcham's synthetic classic book traces the development of the question in a cross-disciplinary investigation, structured by the distinction between "Engineering Philosophy of Technology" and "Humanities Philosophy of Technology" (Mitcham 1994). The landscape is explored in detail and one possible future emerges clearly for Mitcham: "This book may be read as the prolegomenon to inevitably more explicit ethical reflections on technology" (Mitcham 1994: 7; see also Davis 1991; Cutcliffe 2000; Cavalier 2005).

In the wake of Mitcham, Heikkerö's recent and synthetic presentation of the field (Heikkerö 2012) gives an idea of the richness and coherence to be found in ethics of technology, justifying its claim to be a next step in philosophy of technology. The importance recently taken by the main regional approaches (bioethics, environmental ethics, engineering ethics) and by the central questions (control and decision, justice, the good life), as retraced by Heikkerö, gives weight to the central question: "How adequate as tools are the traditional Western ethical concepts and theories in the late-twentieth-century and early-twenty-first-century world?" (Heikkerö 2012: 26). Relying on sources that will be used abundantly in my argumentation, such as Thoreau, Gandhi, Heidegger, and Asian philosophies, Heikkerö transposes this question into a task ("to reveal where ethical questions have been ignored, suppressed, or covered over by technical questions," Heikkerö 2012: 189) and shows that this task calls into question our way of life and our vision of the world: "To put it in simple terms, it seems that humans in advanced societies should find a new way of willing, in addition to or instead of the manipulative willing" (Heikkerö 2012: 193).

The emerging field of technoethics is the subject of the introductory chapter of an extensive textbook (Luppicini and Adell 2008: 1–19) that displays a very broad range of concerns and applications for the discipline. In specific fields, such as cyberethics for instance, the issues are inevitable contemporary concerns – the "moral obligation to bridge the digital divide," "why is spam morally objectionable?" and so on (Tavani 2004: 305, 278). Charles Ess's ethics for digital media (Ess 2009) is another testimony of the growing importance of technoethics in the field of technological media studies and the new spirit it brings forth.

The present research about technology-mediated morality refers to an ambitious philosophical endeavor: "rethink the status of both objects and subjects in moral theory in order to do justice to the hybrid character of human-technology association" (Verbeek 2011: 17). My version, the technoethics of ordinary life, pays considerable attention to ethical humility in this ambition: "Instead of making ethics a border guard that decides to what extent technological objects may be allowed to enter the world of human subjects, ethics should be directed toward the 'quality of the interaction' between humans and technology" (Verbeek 2011: 156).

The Agency of Artifacts

Who or what is acting in technology? Where exactly is the *agency* of contemporary technology? The best answer to this question would certainly be: nowhere and everywhere. The academic literature on agency tends to be less provocative than that, but it is innovative with the very notion of an *agency of artifacts*. Humans are not the only agents and they are not the only ethically considerable entity. Tools and devices have a delegate agency, through human intentionality. The knife has no agency of its own but the human intentionality of cutting off a slice of bread is delegated to the knife. Then it becomes understandable that available tools and devices influence human intentionality and projects because they shape possible action, they frame possible scenarios, possible states of the world, and in the end possible worlds. In one possible world, food is eaten with fingers and teeth; in another possible world knives and forks afford a mediation. In certain cases, it appears that tools and devices shape human action to the point of bearing a shared or hybrid agency: the train does not only make it "possible" for the passengers to follow this precise itinerary and timing. It makes more sense to consider the train, with its passengers and all the technical environment of the railways, as a collective agent with multiple delegate agencies located in human and nonhuman actors. In certain cases, such as the speed bump (to slow down vehicles), artifacts have their own and independent agency. Verbeek (2005) is a valuable introduction to hybrid agency and gives a clear idea of Latour's approach, foundational at least for the vocabulary and standard examples it has introduced. The cases of (apparently) pure artifact agency are a question for the ontology of technology in information ethics in particular (Floridi 2004; Van den Hoven and Weckert 2008). The theory of hybrid agency is of a more universal significance and its ethics is a real challenge. "Technologies affect our actions not just by altering the course of action (like billiard balls do to each other) but by mediating our reasons or motives to act in a particular way," say Swierstra and Waelbers (2010). It means that technologies share some part of the motivational process in humans. This constitutes an internal or psychological form of hybrid agency, to be added to the more visible external and material shared agency – the hybrid system car/driver for instance, where the car by its information, display, and regulating systems is an actor in conjunction with the driver – see Verbeek and Slob (2011) for more case studies.

Latour's theory of the "missing masses" of morality (Bijker and Law 1992) offers clear examples of the agency delegated to artifacts: not buckling the seat belt in a car rings an alarm; speed bumps prevent car drivers from speeding; some European hotel keys, massive and heavy (before they became digital cards), incite the user to leave them at the reception desk when temporarily leaving. Latour's point is that we add to global morality by this ethical delegation. The idea of "heterogeneous networks that bring together actants of all types and sizes, whether human or nonhuman" (Latour in Bijker and Law 1992: 206) is brilliant and for technoethics it is a foundational asset, but there is more to it. Morality delegated to artifacts in this model comes entirely from the designer of technology, from the designer's script or scenario. Moral delegations take place in a domination process, "devices installed by designers to control the moral behavior of their users" (Akrich in Bijker and Law 1992: 216). The real agent is behind the device and it is not the device. The moral agency of the artifact is a means in a domination relation between humans. This limitation is an aspect of a larger critique of the "social construction of artifacts" approach, as it has been formulated by Winner in his "black box" paper (Winner 1993). Winner firmly contests this sociological approach as giving "an account of politics and society that is implicitly conservative, an account that attends to the needs and machinations of the powerful as if they were all that mattered" (Winner 1993: 369) and as being the all too formal instantiation of a global hypothesis: "All the emphasis is focused upon specific cases and how they illuminate a standard, often repeated hypothesis, namely, that technologies are socially constructed" (Winner 1993: 373).

The idea that artifacts are moral agents and not only moral patients has an ontological import that I do not wish to minimize, for it leads to fascinating philosophical novelties such as the possibility of a moral agency without the usual attributes of mental states, freedom and responsibility, particularly concerning informational systems (Floridi and Sanders 2004). My intention is to help transform the largely speculative question of the agency of artifacts into a more pragmatic approach to hybrid agency, which would differentiate and connect a series of notions: affordance, usability, actual uses, innovative uses, and finally the community-driven and user-driven evolution of technology.

The Pragmatic Ethical Turn

The intention of this book is to return to concrete issues, beyond the well-known discourse of technophilia, which is so natural in engineering philosophy, and the equally well-known plea for technophobia, which is so frequent in the humanities and social critique. The question is not about optimism versus pessimism. It is not a debate between utopian narratives (issued by the advertisement industry) and dystopian narratives (issued by the entertainment and infotainment industry). Philosophy of technology has taken "an empirical turn that might roughly be characterized as constructivist" (Achterhuis 2001: 6). I suggest carrying on with a *pragmatic ethical turn* to keep at bay any determinism and a priori valuation of technology, utopian as well as dystopian. Pragmatism helps surmount the persistent resistance to direct ethical questioning in mainstream philosophy of technology (Keulartz 2002).

Philosophers need to seriously address the fluidity of a smartphone interface, the efficiency of a Dyson vacuum cleaner, or the familiar noise of an antique vacuum cleaner. Beyond their phenomenological description, and out of it, the emotional experience acquires moral significance and in some cases even supplies ethical resources for the self. If we leave this dimension of modern experience unaddressed, we may miss something of value in contemporary life. We might miss a specific level of values in the technosphere that we are inhabiting, half-consciously most of the time. Typical issues like the moral value of cycling, of daily physical exercise, of open source software, the moral acceptability of psychotropic medicine or the moral implication of one's driving style, and so many others, deserve a substantial approach from the point of view of philosophical ethics. They are not solely lifestyle options with possible consequences on the community or the ecosphere. Actually, these issues deserve more than an ethical approach; they call for norms and engagements; they need to be part of a mindful moral project. Lifestyle matters; lifestyle prominently matters in contemporary applied ethics.

Fundamental ethical questions are typically non-intuitive and frequently very complex and context-sensitive. These questions refer to universals in human cultures: they are present in every culture, sometimes with an intriguing disparity, but pointing to the hypothesis of a universal moral development scheme (Kohlberg 1981). Our technological civilization addresses these universal ethical questions (such as happiness, health, duty, justice) in its own way, that is to say most of the time only as consumer needs to be catered for by the technological production and marketing systems. But this new way of dealing with essential human questions is poorly correlated with the cultural heritage, the constituting values of the society. This conflict is at the origin of the reemergence of a series of forgotten fundamental (Greek) moral notions: "the good life," which is largely the question of *eudaimonia* ("happiness," how to live and live well), *aretē* ("excellence," chiefly moral excellence), and *wisdom* (not intellectual wisdom, *sophia*, but practical wisdom, *phronēsis*) – see Higgs et al. 2000.

Embedded Values in Technology

The fact that we are not conscious or fully aware of the values system immanent in technology does not mean that there is no such values system. Nor does it mean that it is impossible to gain a clear conscience of this values system. Computer code, for instance, is strongly value-laden (Berry 2011: 9–10) and it is crucial to be aware of what *ethical* values lie behind software and websites.

A good reason to bring this implicit ethics to light is to detect at its source a possible hidden ideology and hidden agenda that would be embedded in technology, or, in a more moderate version of the same danger: it obviously makes sense in ethics to avoid having a blind spot right at the center of one's concern. Unconscious metaphysics and values systems are the worst suppressors of ethical awareness.

The core natural value in technology is *efficiency*. I take efficiency in the somewhat elaborated version in use in the literature:

> Efficiency, through control, describes actions that are more than simply effective [...]. The power of efficiency lies not in producing a great effect, but in producing a desired effect using precisely the desired amount of resources. Effectiveness, in contrast, may often be achieved by dumping resources into a problem, by using more resources than a desired outcome might in fact require if the problem could be specified with precision (Mitcham 1994: 226–227).
>
> (Alexander in Meijers 2009: 1011)

Alexander's monograph on efficiency as a *mantra* shows "how an obscure philosophical concept – and that is how efficiency started out – became an important industrial tool, and, later, a popular social and personal ideal" (Alexander 2008: 2).

This notion of efficiency is cognate with *instrumental rationality*. The pattern of instrumental rationality is the logical pair *means* and *ends*. The philosophical tradition agrees to distinguish (1) a rational assessment of the ends, objectives, goals, purposes, of the action, which raises questions of value, from (2) the mere technical detection, invention, and implementation of the means for theses ends, which is instrumental rationality. Human reason faces two different challenges. One is ethical (discovery of ends), the other one is technical (discovery of means).

The various forms of technological efficiency, in spite of their ambivalence (Kleinman 2009), converge into the broader project of *control*. Borgmann's existential analysis did not miss this point: "Yet, in one sense,

Table 1.1 Apparent Values in Technology (Provisional)

1 Foreground values
   a Comfort (well-being and reassurance)
   b Facilitation
   c Production and consumption (economic growth)

2 Background values
   a Power as a self-sufficient value
   b Control as a self-sufficient value

3 Emergent values
   a Responsibility and precaution
   b Sustainability (ecological, economical)
   c Global justice

4 Potential values
   a Empowerment and autonomy
   b Collaboration and networking
   c Satiety and frugality
   d Care and self-care

technology is nothing but the systematic effort to get everything under control" (Borgmann 1984: 14). There is an ideology of control that "colonizes" Western minds (Hershock 1999). There is an important difference between the direct power of a simple *tool* on reality, and the entire interpretation of the world, the values system that is embedded in a technology or technological system (Hershock 1999: 20–24). A technology is an interpretative system of the world and of possible action in it. Technologies are valuations and in Western technologies control is the basis of this valuation system.

When the focus is on the user-end and not (only) the designer-end, another set of core values emerges: comfort, facilitation, well-being. These are the ends of instrumental rationality and engineering in the contemporary world. These values characterize our form of life (Heidegger 1954). Our basic existential orientation is the exploitation of availabilities and affordances, for our comfort, taken in the most extensive meaning of the notion.

A first draft of the immanent and apparent values system of contemporary technology, at this stage, gives a better idea of what is at stake. The items and categories of Table 1.1 result quite easily from the existing literature in philosophy of technology.

1.2 Ethical Paradigms and the Question of Technology

After this exploration of the problem from the perspective of philosophy of technology we need to configure it from the perspective of ethical theory. I will argue that the challenge of contemporary technology is a shock test for contemporary ethical doctrines and methods, and that few can withstand it successfully.

The Twentieth-century Crisis in Moral Philosophy

During the second half of the twentieth century, a devastating critique ruined moral philosophy from the inside. The rebirth of a viable discipline after this episode is still a matter of discussion. The attack was launched from Great Britain in three waves (Anscombe 1958; MacIntyre 1981; Williams 1985). The main philosophical consequences of this confrontation are still animating the academic conversation. As a result, the present doctrines in moral philosophy are hardly audible. The question of technology acts as an aggravating factor in this crisis but it also offers a promising opportunity for a reconstruction program.

MacIntyre's analysis of the crisis fits best with my approach to technoethics because of his *virtue ethics* orientation (1.3 below). He underlines the lack of meaning for "the key expressions of morality" (MacIntyre 1984: 2), which comes from the lack of a reference values system, after the theological worldview collapsed in modern times. We are still using the language of morals but it relies on nothing. My approach to this crisis is a humanist alternative, acknowledging the irremediable absence of revealed values from a transcendent source (theological or mystical models) and not trying to elaborate a replacement Munchhausen-like auto-foundational set of duties (Baron Munchhausen pulled himself out of a swamp by his hair, or bootstraps).

From the moment ethics does not rest on a preexisting set of values – objectively existing and subjectively mandatory – it can reclaim the status of a quest, a philosophical search, neither for knowledge nor for power, but for a form of practical wisdom, a guidance for human action, individual and collective, born out of rational analysis, plus emotion, plus conversation, plus trial and error.

Applied Ethics

The domain called "applied ethics" has already significantly advanced this reform. An academic architecture has become consensual and it has installed an articulation of issues in three stages:

1   *Meta-ethics* deals with the nature and methods of ethics in general and it tackles the more fundamental and critical questions raised by the crisis.
2   *Normative ethics* takes up the challenge of morality in its traditional intention: statements about what is good and bad, right and wrong, and other moral norms.
3   *Applied ethics* configures the various domains of human activity (medicine, business, war, parenting, and so on) according to the formal options offered by (1) and the material norms offered by (2).

In the best possible scenario, an applied ethics research project is a test and a proof for meta-ethical and normative conceptions, or, even better, it includes "bottom-up" considerations to rebuild some norms and some methods. This scenario has the advantage of avoiding a heavy (meta-ethical) prejudice: the deductive nature of moral reasoning, as opposed to inductive. Technoethics, and particularly ordinary technoethics as presented here, clearly stresses the inductive and pragmatic procedures in ethics.

The second major articulation of ethical philosophy is ternary again, depending on three meta-ethical options:

1   *deontology:* the basic question is about the duties and moral obliga-tions, so it bears on *the motives of the agent*;
2   *consequentialism:* the basic question is about *the consequences of the action* in terms of happiness, or utility under a given definition, of the person and/or a community (more or less extended);
3   *virtue ethics:* the basic question is about *the nature of the agent*, his/her excellence or virtue (Greek *aretē*), with a speculative aspect (what is this excellence?) and a practical aspect (how to achieve it?).

As soon as one tries to use these distinctions, one finds that almost every real-life application to an ethical situation overlaps or blurs the defined categories. To put it differently, almost any given moral stance can be for-mulated in the three idioms – deontology, consequentialism, virtue ethics. Whereas inventing fictional cases where the three categories clash is a favored activity in meta-ethics it does not provide as yet a reason to opt for one method in particular in applied ethics.

Virtue ethics is gaining in influence (after having suffered the highest casualties from the crisis) and I will argue that it gives a better ground for technoethics, at least in the "ordinary" version that I offer here.

Pragmatism in Ethics

The dominating paradigm in applied ethics comes from medical ethics and can be called "principlism." Its outcome is a list of principles, correspond-ing to values such as autonomy, beneficence, and justice (Beauchamp and Childress 1979). These are obviously important values for any ethics but the following question remains: are deontological *principles* the right place to start from? The discursive orientation of mainstream applied ethics is convenient for writing ethical charts and substantiating ethical institutional procedures. If the balance of power slowly shifted toward the ethical empowerment of individuals, versus institutions, as I will argue, and if at the same time the societal conversation became more about values and projects than about rules and regulations, then values such as autonomy, beneficence, and justice could be considered as *ends*, more than principles. These values would be enlisted as ends, as objectives, in the project of indi-vidual lives and human communities.

Shifting from questions of principles to questions of ends means shift-ing from a deontological and discursive orientation in ethics toward a pragmatic and personal (virtue ethics) orientation. In this perspective, morality is not preeminently the statement of rules that people have to comply with. Instead, it can be the search for happiness as the universal natural end (utilitarian ethics), or the quest for self-improvement as a final end (virtue ethics), or even a broader acceptation of flourishing as an end (encompassing the flourishing of nonhuman entities). In the ethics of technology, the privileging of "principles" is a symptom of technology being assessed from the designer-end point of view, or more precisely from the dominant point of view, the place where rules are decided and enacted. The opposite vision of technology, favored in this book, is centered on everyday use and microactions. It is not charts and principles to be applied to ordinary life that is called for, but rather ends, meanings, and projects that would illuminate micro-choices and ordinary behaviors.

More than ethical expertise to devise principles and to apply them in case studies, *technology as the texture of our everyday life* demands practical guidance. This situation suggests a more pragmatic approach than "principle ethics." In ordinary parlance, when people say "In principle, we should..." they mean that the principle does not help much. Technoethics directly aims at the pragmatic decision level because it takes technology primarily as the substrate of everyday life. Technoethical issues emerge from this existential level and they are treatable within this existential level.

Some methods in applied ethics are openly pragmatic in the sense I am referring to. McGee's framing of the "perfect baby" problem in genetics can be an example of those (McGee 1997). This kind of pragmatism has an early source in ethics with Sidgwick's classic survey of ethical motivation sources, affirming that

> there are certain absolute practical principles, the truth of which, when they are explicitly stated, is manifest; but they are of too abstract a nature, and too universal in their scope, to enable us to ascertain by immediate application of them what we ought to do in any particular case; particular duties have still to be determined by some other method.
>
> (Sidgwick 1907: III, 13, 3)

To discover and expound these "other methods" academic ethics has long been paralyzed by a supposed Kantian veto on moral psychology – no source of morality can be psychological, i.e., empirical. After these restrictions were overthrown by Jean Piaget's constructionist psychology and similar research, empirical studies of human psychological development became possible. Lawrence Kohlberg's model of the stages of moral development in particular deserves a serious consideration by moral philosophers (Kohlberg 1981). It suggests a progressive scale of moral motivation:

1   moral insensibility (non-sentient entities);
2   reward/punishment as motivation (animal, very young human);

3   group integration as motivation (children, adolescents);
4   universal rules, understood and accepted, and self-accomplishment as motivation (moral maturity).

This fourth stage may be taken as an achievement in applied ethics. Moral maturity as the end and as the intrinsic value in the flourishing of a human person meets the criteria of a consensual value in today's philosophical landscape. To make it really consensual, I will insist as strongly as Kohlberg himself on the difference between moral education and ideological indoctrination in the self-construction of a human person. This standard induces a humanist applied ethics, relying on natural (but cultural) ethical skills. They lead to a moral equilibrium acquired through a personal history of (guided) personal experiences and similar to the maturation of logical skills according to Piaget.

The revival of ethics occurs as a form of return to the Greek past. Particularly important is the return to the original notion of *ethos* in Aristotle. In Aristotle's *Nicomachean Ethics* a notion of *ethos* meaning "effective behavior" (as in "ethology") is active. It supposes that the word *ethic* is derived from the Greek *ethos*, meaning "custom," "habit," and "deliberate personal behavior" (facts and actions); not from the Greek *ēthos*, meaning "moral character" (a disposition, a virtuality). Independently of the relationship between the two notions (*ethos* and *ēthos*), the methodological option I want to highlight is the focus on moral action as a fact occurring in the world instead of a focus on moral judgment as a mental process or a causal reconstruction.

Ancient moral schools paid particular attention to the *akrasia* question: the situation in which the agents know what they must do, are willing to do so, but just cannot. Here moral action is separated from moral judgment. A pragmatic method in contemporary ethics gives up the upstream analysis of moral judgment in order to seriously weigh the material actions of agents, particularly their habitual and deliberate small actions, the real substance of a life and the real target for an applied ethics. Therefore, technoethics is an action ethics, largely invested in empirical case studies. Hence, it is quite different from the "principle" applied ethics now prominent in bioethics, environmental ethics, or business ethics.

Ethics vs. Politics

I suggest for technoethics a *post-political* path. This post-political will often turn out to be simply *micro-political* and in this sense to be the real political. I think Winner's expression remains entirely true: "*technē* is *politeia*," our *politeia* (Winner 1986). But *politeia* can be split into three social realities that may be divergent in the contemporary context:

1   *The social level*, where the actors are human communities in their diversity, from workers unions to business corporations, from militant

associations to utility organizations, not forgetting the media. This level is the object of sociology.

2 *The institutional level*, where the actors are typically bureaucratic collective agents, differentiated by their official status from any other social actor. Governments and all their agencies are the traditional level of *institutional politics*. This level cannot stand as the only political level.

3 *The societal level*, where the actors are the informal communities of human persons. The term "societal" owes its success to its unique capacity to separate the "real" society of humans from its allegedly "representative" institutions (politicians and other social representatives). The agency here belongs to networks of human persons. Social movements related to the Internet, but also the public opinion on bioethical issues, are clear instances of a distinct societal level.

The complex and largely unknown interactions between these three levels are an important theoretical question. Technoethics addresses concrete cases on the societal level and directly takes them as ethical questions – personal as well as collective. My hypothesis is that this method allows an innovative approach to the question of power. The investigation is still about power, but not institutional power. I take Winner's axiom about *technē/politeia* not in his sense: "our instruments are institutions in the making" (Winner 1986: 54), but in this post-political and ethical sense: our instruments are lifeforms in the making – and we can have more agency and more responsibility in this making, as personal selves or as collaborative communities. Winner's relative pessimism can be superseded. I agree that artifacts have politics, they shape reality, and possibly according to a secret agenda. This is a pressing concern for democracy and an urging preoccupation in some cases – the acceptability of nuclear energy or global surveillance for instance. But these problems are linked to the very poor standards for technology assessment and social acceptation. Assessment institutions are more than often poor enough to be mocked, as in the chapter "Brandy, cigars and human values" (Winner 1986: Chapter 9). Fake, tainted, and corrupted assessments are now the object of book-length exposures (Oreskes and Conway 2010). I try in this book to imagine how civic virtues and behaviors can "hack" technology assessment and social acceptation.

Technoethics vs. Patronization

The danger in a shift to ethics is obviously moralization as patronization: the arrogant lessons given from a self-proclaimed moral superiority. There is a danger of being ridiculed but also and more importantly of being counterproductive. Taking the option of a "virtue" ethics makes this reproach even more probable, on the surface. But the recent context of applied ethics provides *non-moralizing evaluation* methods (Slote 1992). A modest use for ethical theories can be found: "Theories would serve not as authoritative major premises in some deductive syllogism, but rather as deliberative tools, drawing attention to aspects of the case to which attention should be given and that might otherwise be ignored or overlooked" (Chadwick 1998: 188). In Chadwick's vision, "ethical expertise" is a procedural rather than a substantive skill. The characteristics of the ethical turn in philosophy of technology strongly put forward this method (Swierstra 1997):

1   a *constructive* focus shift, from survival to the good life: away from dystopian prophecy;
2   acceptation of the *naturality* of technology in human life: away from anti-artificialism;
3   renewal of *individual responsibility*: away from social determinism.

These features are disruptive enough in the context of a predominantly technophobic and sociological (old style) philosophy of technology and engineering. Technoethics emerges as an academic field that remains in connection with philosophy of technology but is more and more self-sustaining in its questions and methods. Along with the *International Journal of Technoethics*, a bulky *Handbook* (Lupiccini and Adell 2008) open-mindedly delineates the field. The table of key areas in technoethics provided in this volume displays the interdisciplinary vocation of technoethics, possibly as a uniting factor in the ethical paradigms of today. The list is as follows: Computer Ethics, Engineering Ethics, Internet Ethics and Cyberethics, Media and Communication Technoethics, Professional Technoethics, Educational Technoethics, Biotech Ethics, Environmental Technoethics, Nanoethics, Military Technoethics (Lupiccini and Adell 2008: 15).

1.3 Methods for Technoethics in the Ordinary

This book will answer questions like these: How to resist the perfect drug, and why? Why is social networking online most of the time a shallow and fake experience? Why do some people love their car? Why do we need digital connections just as much as we need human connections, and is it morally acceptable? My questions are thus not only centered on technoethics, but on ordinary technoethics. The argument to justify this focus is that the ordinary technosphere is the proximal technosphere, our permanent interface with the world, our lifeworld. My second key methodological choice is to favor a constructive and non-confrontational method. Technoethics aims at inspiring action, ordinary action.

Practical Wisdom vs. Sterile Scholasticism

> These days, however, there is little connection between the teaching
> and publishing of professional philosophers, on the one hand, and
> anyone's effort to actually become better, on the other. Even if the self-
> help genre is mostly trite, at least it professes to aim at changing peo-
> ple's life for the better; few philosophical works claim that.
>
> (Angle 2009: 160)

My emphasis is on practical wisdom as opposed to what I will call "scholasticism" in contemporary philosophy – I use "scholasticism" in the sense
found in Mitcham (1994) or Rosen (2012). Why practical wisdom instead
of scholasticism? Is the crisis in the humanities severe to the point of disrupting the classical pattern "good guidance comes from good theory"? I
think it is, in some disciplines at least, and technoethics is one. A question
about the compatibility between a strong version of A-ism and specially
devised unreal petty case stories of B-ism is a question of scholasticism,
fascinating as a chess game, and academically fascinating too because there
will never be an end to the flow of publications and rejoinders between
A-ism and B-ism, so professionals and departments will survive. Or not. I
have a feeling that the societal acceptability of scholasticism is becoming
weaker and weaker as the need for substantial and solid ethical guidance is
getting stronger and stronger. I take for granted, in my opting for a reno-
vated virtue ethics, the critique of formalism in ethics (Stocker 1976;
Louden 1984; Midgley 1989) and in particular their disappointment in
face of the impoverishment of moral analysis by scholasticism. I embrace
virtue ethics' advice to advance an agent-centered and not act-centered
ethics (Crisp and Slote 1997). I adhere to a humanist goal for ethics, which
is clearly not to classify actions but to help human persons to grow, i.e., to
assist them in conceiving and realizing human projects. For this task, most
of the existing ethical theories are inadequate because of their formal
reductionism (Pincoffs 1986; Appiah 2008). Our daily intensive and transparent relationship with technology is not a secondary field where questions can be settled by simply applying and refining the principles that have
been set beforehand.

This method resists social sciences' sterile imitation of the supposed
"exact sciences," which is in practice an ideological reduction to *the
command-and-control* model, as Borgmann signaled:

> The professed paradigm of many mainstream social scientists in this
> country is a set of laws that would rival the laws of natural science in
> precision and predictive power. Why has this paradigm in spite of its
> clear if not conclusive failures of precision or applicability not been
> abandoned? We may conjecture that the work of its adherents is sus-
> tained by another, implicit paradigm, that of getting social reality
> under technological control.
>
> (Borgmann 1984: 75)

A Quick View of the Ordinary

Why focus on the "ordinary" and what is it exactly? I just want to point out for the moment that mine is a deflationist approach to moral issues, quite different from the rhetoric of absolutes (absolute danger and grandiose moral salvation) that reigned for a time in philosophy of technology. Technoethics certainly began with the awe of the first nuclear weapons and the critique of the military–industrial complex behind it. It can still be prompted by nuclear energy issues and the financial-industrial complex behind it, and also by some other large-scale extraordinary threats. Yet I will argue that our biggest global structures are driven by the microactions of ordinary citizens doing ordinary things such as traveling, eating, texting a friend, taking a shower, and surfing the Web. And the inaugural analysis of modern existence, in Heidegger's work as well as its application to contemporary technology in Borgmann's work, have revealed this first ethical principle: our dereliction lies in the ordinary. If the ethical status of humankind were a consequence of transcendent extraordinary affairs, such as original sin and redemption for example, nothing important would be attached to what is in one's diet or how one warms one's place (provided no taboo is broken). But with moral predicaments such as climate, energy, digital literacy, health justice, and so on, the battlefield is on the ground of ordinary decisions and actions. In this context, it makes no sense to lecture from above, from a morally dominating abstract stance. Humility is a constitutive virtue of wisdom in my conceptual frame (Table 1.4). The idea is to delve into the ordinary use of technology, to stick to ordinary intimacy with artifacts and technosystems, and to reemerge with actionable guidance for the individual agent.

A sort of checklist of ordinary technology in its diversity can be sketched here (see Table 1.2) and it can be referred to at any stage of the following analysis. Every item in the list is mentioned somewhere in the book. Everyone will have one's own list, in terms of items and of priority order.

Inspirations and Methods

How can philosophy dare claim it could provide guidance and even normativity in the ordinary affairs of life? By accepting engagement, in method as well as in substance, but with a clear conscience of the dangers involved: arrogance and patronization. As a matter of fact, the technoethics I offer makes it a priority to dismiss technocratic arrogance and philosophical patronizing. This challenge is part of the idea of wisdom ethics. I take wisdom to be the most general answer to the reemerging question of the good life. This question today calls for incentives to awareness, through conversation and education, but always leaving choice and responsibility to the human individual person. Technoethics construed as a sapiential question is neither a matter of knowledge, calling for experts, nor a matter of ideology, calling for a guru.

Table 1.2 Ordinary Technology Domains and Instances (a simple list)

Domain	Instances
Automobile	Owning a car, for a household – a dent on the car's side – driving style
Personal computer	Tablet vs. laptop – buy a new computer – ordinary coding: HTML and macros – illegally duplicated software
Internet (Web, email, Skype, social networking sites, e-commerce, etc.)	POP or webmail – Twitter – LinkedIn or equivalent professional networks – comment on a restaurant on Google – comment about a seller on a peer-to-peer commercial website – check and order on Amazon – using three different browsers to bypass the limited number of articles in free access on several information websites – spam
Telephone and smartphones	Buy a new mobile phone – "quantified self" movement: specific tools and sharing practices vs. autonomous practice – e-health applications
Energy (consumption and production)	Solar panel and windmills for personal production vs. grid dependency – awareness of one's consumption
Transportation (other than personal car)	Public transportation systems – online taxi-replacement
Money, payment card	Shopping offline and online
Cooking, eating, drinking	Cooking from raw ingredients – processed food and beverage – drinking sugared sodas – the electric kitchen mixer – devices that allow one to process things by oneself instead of buying processed products
Domestic comfort and appliances	Hot shower – central heating and air conditioning – high pressure washer
Electronic leisure (TV, music, games, etc)	The hifi experience – video games – (electronic) musical instruments – electronic guitar tuner
Personal gear	Washing, cleaning (clothes and personal objects) – how many shoes? – the replacement policy of clothes – carrying a knife
Self-care	Body hygiene and care in general – exercising and sport – going for a walk in nature (instead of TV watching) – being a TV fan of sport
Medicine	Medication (chemical) and biological tests – psychotropic medication
Addictive harmful substances	Smoking tobacco – managing one's alcohol intake

The combination of three main inspirations, namely European humanism, Anglophone pragmatism (mainly in philosophy of technology), and Asian traditions will bestow originality on the technoethics suggested here.

Practical wisdom can be traced back to Aristotle's *phronēsis*, in Latin *prudentia*, the form of "practical wisdom" (different from *sophia*, "theoretical wisdom") elaborated in his *Nicomachean Ethics*. Reconnecting with this foundational moment of ethics, technoethics aims at offering a prudential/sapiential alternative to the "command-and-control" instrumental rationality that characterizes modern technology. Aristotle insisted on the primacy of *praxis* in ethics: virtue is acquired in practicing virtue, in general and for every particular virtue, exactly as in *technē* for a carpenter or a musician.

The pragmatic turn in philosophy of technology corresponds to the pragmatic turn in ethics. A confluence of Dewey and Sidgwick appears in some shared insights, in particular the insight that values are projects and, more specifically, personal projects. Values are defined and selected by definite persons in definite contexts, then put on trial, that is to say implemented and assessed through a living process in the real world. Table 1.1 ("Apparent Values in Technology") evokes a full range of emotions, not necessarily the passionate emotions usually associated with moral situations, but the modest emotions of well-being and everyday easiness. Practical wisdom must take into account the emotional context of our relation to artifacts: euphoric confidence when driving a car, uneasiness with a computer's new operating system, deep memories awakened by the smell of cooking or laundry. The black and white vision of technoethical situations in the academy is often lacking in context, especially because it ignores the emotional context.

New trends in moral philosophy highlight the question of the good life in a manner that proves instrumental to technoethics. A central reference, Tiberius (2008) offers a "realistic optimism" (Tiberius 2008: 152) and a pragmatic method (Tiberius 2008: Chapter 6) that do not lose the reader in the sands of arbitrary abstractions. Instead, they provide tools to deal with ordinary moral tensions and to help the ordinary human person. An "empirically informed ethics" (Tiberius 2008: 21) shuns the ethics of pure rational control and rehabilitates the role of emotions (Tiberius 2015: 219).

To complete this innovative method, a systematic dewesternization of philosophical questions is called for. Several inspiring works offer an approach to the predicament of modernity from the point of view of Asian thought – essentially from Buddhism (Hershock 1999; Hershock et al. 2003), from Daoism (Bynum 2006), and from Neo-Confucianism (Angle 2009).

An Ethics of Wisdom

A specific ethics of virtue, applied to ordinary technology, enlarges the span of analysis: in ordinary life emotions are involved as well as knowledge; routine behavioral habits as well as well-thought decisions; every affective bond, seen and unseen, acknowledged or not. The interface between human and world, or human and artifact, always takes place in a personal history with its own narrative, its own framework of values and goals, its own potentials for flourishing and harmony. Moral wisdom as conceived by virtue ethicists explicitly integrates the important notion of harmony, internal and external. In every example or illustration, the most diverse variety of human flourishing styles is acceptable.

For these reasons, technoethics can only be a virtue ethics: a practice of the self, designing and using resources for its own flourishing, in harmony with an environment – which is not only nature, but also the human environment and the technological environment. Ordinary technoethics is meant to be the central part of what is called after Michel Foucault's last works the *practice of the self*. The practice of the self in a technological world is partly composed of direct "technologies of the self," another Foucauldian notion to be integrated into technoethics, but not in the crude form now promoted by transhumanism. A humanist version is possible for the techniques and technologies of the self in response to the ethical and practical challenges involved in the project of becoming a human person in a world where technology is pervasive. The basic value is the person's life project. From there, self-building is conceived as the dignifying human activity by excellence – this trait can be taken as a short definition of humanism.

The practices of the self will be methodologically dispatched and considered as *microactions*. Microactions are actions, as opposed to discourse. Microactions are ordinary, humble, unseen actions, as opposed to heroic and iconic deeds. One of the main tenets of this book will be that macro-questions, which have remained unsolved in modernity, require a change of setting: from the abstract global level to the concrete local level, the latter being the uncharted territory of ordinary technoethics. Shantideva used an expressive image in the eighth century, which was used again later by the Transcendentalists (Emerson 1983: 994; Thoreau 2004: 46):

> Where could I possibly find the leather
> To cover with leather the whole surface of the earth?
> But with leather just on the soles of my shoes,
> It's the same as having covered the entire earth's surface.
>
> (Shantideva 763: 5/13)

The leather on the sole of our shoes is the ordinary wisdom that potentially transforms the whole technosphere into an inhabitable lifeworld for a flourishing human person. Technoethics contributes to this harmony as a wisdom ethics. Wisdom is one of the unaddressed affordances of the technosphere.

The present state of the debate might well be an inflection point in the philosophy of values (ethical, social, political), a moment where the insistence must shift from *power* to *wisdom*, to put it in its simplest form. In the West, from Plato's time on, human existence has been assessed by philosophers from the command-and-control point of view: the real project is power and the justification for power is the good life – which has a different sense for the dominant and the dominated, a clause usually non-specified. The time has come to consider the wisdom point of view, on the personal and on the collective level: the real project is the good life – here the good life in a personalized sense for every person and every community, a clause necessarily specified. Ordinary technoethics cannot be separated from this essential transition, the transition from power to wisdom.

I will adopt a triple distinction in the concept of power to situate technology, wisdom, and power in a common perspective, as shown in Table 1.3.

Our form of life takes (1) as its first choice. In face of any problem, we try to find a technological solution, if possible a "quick fix" solution. For instance: I have a headache so I need a pill; we want to build a road from A to B so the hill in between will be excavated; some people overseas are a security threat so a killer drone must suppress the threat.

When material technology fails or is unavailable, domination over other humans will be attempted. This option (2) of power is often equivalent to a political technology: an instrumental rationality that ends in the practical question of identifying the right person or institution to submit to. Instances are the cases of any problem X (economical, social, environmental, humanitarian) that is automatically translated into these terms: who or what do we need to submit to in order to solve problem X? Domination by leaders and institutions of all kinds is the second form of power that supplements and assists technology.

Option (3) is the last one that would occur to us in ordinary life, sadly enough. The reason is easily found in the effort that power over oneself implies. When something cannot be provided by technology, and cannot even be made available by an institution of power, one must act by oneself and sometimes act on oneself. Wisdom eminently belongs to this third mode of power, the one we tend to ignore. To restore it and make it the determinant resource of human life we need to reassess the nature and power of technology and to reassess the benefits and costs of domination.

The fact that option (3) is *non-confrontational* will be of importance. Without it, wisdom would be no third option, but just another technology, or just another domination scheme. Ethical philosophy can be limited to an accompaniment of life, personal and collective. This is a modest but resourceful attitude, particularly after a twentieth century of violent ideologies. Non-confrontational Asian doctrines retain many concepts that are already active in tentative ethics of modernity. Stoic philosophy had a notion of "accompanying the events" (Epictetus 1928: III, x, 18).

Table 1.3 The Three Kinds of Power

1	Power over things	Technology
2	Power over others	Domination
3	Power over oneself	Wisdom

Table 1.4 Wisdom (Elementary)

Wisdom fundamental practices
Awareness
Autonomy
Harmony
Humility
Benevolence
Courage

The reappearance of this idea in European philosophy and ethics of technology is remarkable. Hottois characterizes the philosophical accompaniment of contemporary technology as disrupting "the dialectic of foundation, power, and mastery, and promoting a free relationship with neither subordination of technology to symbol nor the opposite subordination" (Hottois 1996: 16). Verbeek's conclusion about "accompanying technology" is exactly in this spirit (Verbeek 2011: 156). Not every project of radical change need be a confrontation.

In order to proceed to pragmatic analysis, I need to suggest one definite version of wisdom, composed of key "virtues," which can be in turn combined with each other to give birth to an open list of virtues. The entire book is built on the development and justification of this framework and its application to the technosphere. The list (as given in Table 1.4) is at this stage hypothetical. As a consequence of the pragmatic orientation of this book, the universal virtue of wisdom is dispatched not in derivative "virtues" but in "fundamental practices." Lists of virtues abound in the moral literature of the past and many of them are inspirations for the list given here. Among them are the virtues list of Mahayana Buddhism – "generosity, virtuous conduct, patience, energetic commitment, meditation, and discernment" (Hershock 2014: 21); Zen esthetic of life – irregularity (*fukinsei*), simplicity (*kanso*), unpretentious naturalness (*shizen*), tranquility (*seijaku*), and freedom from convention (*datsuzoku*)" (Hershock 2014: 125); Neo-Confucian virtues (Angle 2009); and virtues ethics reference lists (van Hooft 2013: Chapter 5).

References

Achterhuis, Hans (ed.). 2001. *American philosophy of technology: The empirical turn.* Trans. R.P. Crease. Bloomington, Ind.: Indiana University Press.

Alexander, Jennifer K. 2008. *The mantra of efficiency: From waterwheel to social control.* Baltimore, Md.: Johns Hopkins University Press.

Angle, Stephen C. 2009. *Sagehood: The contemporary significance of Neo-Confucian philosophy.* Oxford University Press.

Anscombe, G.E.M. 1958. Modern moral philosophy. *Philosophy* 33: 1–19.

Appiah, Kwame Anthony. 2008. *Experiments in ethics.* Cambridge, Mass.: Harvard University Press.

Aristotle. 350 BCE. *Nichomachean ethics.* Trans. W.D. Ross (1925). http://classics.mit.edu/Aristotle/nicomachaen.html.

Baker, Lynne Rudder. 2004. The ontology of artifacts. *Philosophical explorations* 7(2): 99–111.

Beauchamp, Tom L., and James F. Childress. 1979. *Principles of biomedical ethics.* 3rd edn, 1989. Oxford: Oxford University Press.

Berry, David M. 2011. *The philosophy of software: Code and mediation in the digital age.* London: Palgrave Macmillan.

Bijker, Wiebe E., and John Law (eds). 1992. *Shaping technology/Building society: Studies in sociotechnological change.* Cambridge, Mass.: MIT Press.

Böhme, Gernot. 2012. *Invasive technification: Critical essays in the philosophy of technology.* Trans. C. Shingleton. London: Bloomsbury Academic.

Borgmann, Albert. 1984. *Technology and the character of contemporary life: A philosophical inquiry.* University of Chicago Press.

Brey, Philip, Adam Briggle, and Edward Spence (eds). 2012. *The good life in a technological age.* London: Routledge.

Bynum, Terrell Ward. 2006. Flourishing ethics. *Ethics and information technology* 8(4): 157–173.

Cavalier, Robert J. (ed.). 2005. *The impact of the Internet on our moral lives.* Albany, N.Y.: State University of New York Press.

Chadwick, Ruth (ed.). 1998. *Encyclopedia of applied ethics* (4 vols). San Diego, Cal.: Academic Press.

Crisp, Roger, and Michael Slote (eds). 1997. *Virtue ethics.* Oxford: Oxford University Press.

Cutcliffe, Stephen H. 2000. *Ideas, machines, and values: An introduction to Science, Technology, and Society studies.* Lanham, Md.: Rowman & Littlefield.

Davis, Michael. 1991. Thinking like an engineer: The place of a code of ethics in the practice of a profession. *Philosophy and public affairs* 20(2): 1150–1167. www.jstor.org/stable/2265293.

Ellul, Jacques. 1954. *La technique, ou l'enjeu du siècle.* Paris: A. Colin, repr. Economica, 1990. (*The technological society.* Trans. John Wilkinson. New York: Knopf, 1964).

Emerson, Ralph Waldo. 1983. *Essays and lectures.* Ed. J. Porte. New York: The Library of America.

Epictetus. 1928. *Discourses, Manual, Fragments.* Ed. W.A. Oldfather. Cambridge, Mass.: Harvard University Press.

Ess, Charles. 2009. *Digital media ethics.* Cambridge: Polity Press.

Feenberg, Andrew. 1999. *Questioning technology.* London: Routledge.

Feenberg, Andrew. 2002. *Transforming technology: A critical theory revisited.* Oxford: Oxford University Press.

Floridi, Luciano (ed.). 2004. *The Blackwell guide to the philosophy of computing and information.* Oxford: Blackwell.

Floridi, Luciano, and J.W. Sanders. 2004. On the morality of artificial agents. *Minds and machines* 14(3): 349–379. www.philosophyofinformation.net/publications/pdf/omaa.pdf.

Frey, R.G., and Christopher Heath Wellman (eds). 2003. *A companion to applied ethics.* Oxford: Blackwell.

Giddens, Anthony. 1991. *Modernity and self-identity: Self and society in the Late Modern Age.* Cambridge: Polity Press.

Heidegger, Martin. 1927. *Sein und Zeit.* 13. Auflage, 1976. Tübingen: Niemeyer. (*Being and time.* Trans. J. Macquarrie and E. Robinson. London: Harper & Row. 1962).

Heidegger, Martin. 1954. Die Frage nach der Technik. In *Vorträge und Aufsätze*, Pfullingen: Neske. (*The question concerning technology, and other essays*. Trans. W. Lovitt. New York: Harper Torchbooks, 1977).

Heikkerö, Topi. 2012. *Ethics in technology: A philosophical study*. Lanham, Md.: Lexington Books.

Hershock, Peter D. 1999. *Reinventing the wheel: A Buddhist response to the information age*. Albany, N.Y.: State University of New York Press.

Hershock, Peter D. 2014. *Public Zen, personal Zen: A Buddhist introduction*. Lanham, Md.: Rowman & Littlefield.

Hershock, Peter D., M. Stepaniants, and R.T. Ames (eds). 2003. *Technology and cultural values: On the edge of the third millennium*. Honolulu: Hawaii University Press.

Higgs, Eric, Andrew Light, and David Strong (eds). 2000. *Technology and the good life?* Chicago, Ill.: University of Chicago Press.

Hilpinen, Risto. 2011. Artifact. *The Stanford encyclopedia of philosophy* (Winter 2011 Edition). Edward N. Zalta (ed.). http://plato.stanford.edu/archives/win2011/entries/artifact.

Hottois, Gilbert. 1984. *Le signe et la technique. La philosophie à l'épreuve de la technique*. Paris: Aubier.

Hottois, Gilbert. 1996. *Entre symboles et technosciences*. Seyssel: Champ-Vallon.

Ihde, Don. 1979. *Technics and praxis*. Dordrecht: Reidel.

Ihde, Don. 1990. *Technology and the lifeworld: From Garden to Earth*. Bloomington, Ind.: Indiana University Press.

Keulartz, Jozef. 2002. *Pragmatist ethics for a technological culture*. Dordrecht; Boston: Kluwer Academic Publishers.

Kitcher, Philip. 2001. *Science, truth and democracy*. Oxford: Oxford University Press.

Kleinman, Sharon (ed.). 2009. *The culture of efficiency: Technology in everyday life*. New York: Peter Lang.

Kohlberg, Lawrence. 1981. *The philosophy of moral development: Moral stages and the idea of justice*. San Francisco, Cal.: Harper & Row.

Kranzberg, Melvin. 1986. Technology and history: "Kranzberg's Laws." *Technology and culture* 27(3): 544–560.

Louden, Robert B. 1984. On some vices of virtue ethics. *American philosophical quarterly* 21: 227–236.

Luppicini, Rocci, and Rebecca Adell (eds). 2008. *Handbook of research on technoethics* (2 vols). Hershey, Pa.: Information Science Reference.

Macintyre, Alasdair. 1984. *After virtue: A study in moral theory*. 2nd edn. Notre Dame, Ind.: Notre Dame University Press.

Marcuse, Herbert. 1964. *One-dimensional man*. Boston, Mass.: Beacon Press.

McGee, Glenn. 1997. *The perfect baby: A pragmatic approach to genetics*. Lanham, Md.: Rowman & Littlefield.

Meijers, Anthonie (ed). 2009. *Philosophy of technology and engineering sciences (Handbook of the philosophy of science, vol. 9)*. Amsterdam: Elsevier.

Midgley, Mary. 1989. *Wisdom, information and wonder: What is knowledge for?* London: Routledge.

Mitcham, Carl. 1994. *Thinking through technology: The path between engineering and philosophy*. Chicago, Ill.: University of Chicago Press.

Mitcham, Carl. 1997. *Thinking ethics in technology: Hennebach lectures and papers, 1995–1996*. Golden, Colo.: Colorado School of Mines.

Olsen, Jan Kyrre Berg, Stig Andur Pedersen, and Vincent F. Hendricks (eds). 2009. *A companion to the philosophy of technology*. Hoboken, N.J.: Wiley-Blackwell.

Oreskes, Naomi, and Eric Conway. 2010. *Merchants of doubt: How a handful of scientists obscured the truth on issues from tobacco smoke to global warming.* New York: Bloomsbury.

Pincoffs, Edmund L. 1986. *Quandaries and virtues: Against reductivism in ethics.* Lawrence, Kans.: University Press of Kansas.

Pitt, Joseph C. (ed.). 1995. *New directions in the philosophy of technology*. Dordrecht: Kluwer (Philosophy and Technology, vol. 11).

Rosen, Michael. 2012. *Dignity: Its history and meaning*. Cambridge, Mass.: Harvard University Press.

Shantideva. AD 763 *Bodhicaryâvatâra*. www.berzinarchives.com/web/x/pdf/?type=pdf&book=true&path=/web/x/prn/p.html_1487505749.html&__locale=en.

Sidgwick, Henry. 1907. *The methods of ethics*. 7th edn. London: Macmillan. www.laits.utexas.edu/poltheory/sidgwick/me/index.html.

Slote, Michael. 1992. *From morality to virtue*. Oxford: Oxford University Press.

Stocker, Michael. 1976. The schizophrenia of modern ethical theories. *Journal of philosophy* 73: 453–466.

Swierstra, Tsjalling. 1997. From critique to responsibility: The ethical turn in the technology debate. *Techné* (3)1: 4–48.

Swierstra, Tsjalling, and Katinka Waelbers. 2010. Designing a good life: A matrix for the technological mediation of morality. *Science and engineering ethics*, Nov. 2010. doi:10.1007/s11948–010–9251–1.

Tavani, Herman T. 2004. *Ethics and technology: Ethical issues in an age of information and communication technology*. 3d edn. Hoboken, N.J.: John Wiley & Sons.

Thoreau, Henry David. 2004. *The Higher Law. Thoreau on civil disobedience and reform*. Ed. W. Glick. (H.D. Thoreau, *The writings*). Princeton, N.J.: Princeton University Press.

Tiberius, Valerie. 2008. *The reflective life: Living wisely within our limits*. Oxford: Oxford University Press.

Tiberius, Valerie. 2015. *Moral psychology: A contemporary introduction*. London: Routledge.

Tiles, Mary, and Hans Oberdiek. 1995. *Living in a technological culture: Human tools and human values*. London: Routledge.

Toulmin, Stephen. 1990. *Cosmopolis: The hidden agenda of modernity*. New York: The Free Press (Macmillan).

Van den Hoven, Jeroen, and John Weckert (eds). 2008. *Information technology and moral philosophy*. Cambridge: Cambridge University Press.

Van Hooft, Stan. 2013. *Understanding virtue ethics*. Cambridge: Cambridge University Press.

Verbeek, Peter-Paul. 2005. *What things do: Philosophical reflections on technology, agency, and design*. Trans. R.P. Crease. Pennsylvania: Pennsylvania State University Press.

Verbeek, Peter-Paul. 2011. *Moralizing technology: Understanding and designing the morality of things*. Chicago, Ill.: University of Chicago Press.

Verbeek, Peter-Paul, and Adriaan F.L. Slob. (eds). 2011. *User behavior and technology development: Shaping sustainable relations between consumers and technologies*. New York: Springer.

Williams, Bernard. 1985. *Ethics and the limits of philosophy*. London: Collins.

Winner, Langdon. 1977. *Autonomous technology: Technics-out-of-control as a theme in political thought*. Cambridge, Mass.: MIT Press.

Winner, Langdon. 1986. *The whale and the reactor: A search for limits in an age of high technology*. Chicago, Ill.: University of Chicago Press.

Winner, Langdon. 1993. Upon opening the black box and finding it empty: Social constructivism and the philosophy of technology. *Science, technology, and human values* 18(3): 362–378.

2 *Technosapiens*
The Coevolution of Nature, Humankind, and Technology

Humankind cannot hide how different it is from any other life form on the planet, including the original Homo Sapiens and the diverse branches of the genus Homo which are now extinct. Today the unique character of our species is more than a biological fact. We know that our genome is a perfectly ordinary primate genome and that it has not changed since the Paleolithic era. However, the obviously specific form of life deployed by modern humankind is enough to determine a radically different species in terms of its behavior and its very particular integration into the planet's ecosystem. The term "Anthropocene" is now on its way to be validated by geological authorities as the name for this period where the major transformative factor of Earth is human activity (Grinevald 2008). From a philosophical point of view, it makes sense to coin the term "Homo Sapiens Technologicus" (Puech 2008) – or "Technosapiens" in its shorter gender-neutral form. I will argue that we are a new species, not only because of the massive consequences of our activities in the ecosystem, which is a one-way causal and simplistic observation, but because we are agents in a new form of global evolutionary process involving nature, humankind, and technology. Let us call "coevolution" the braid composed by these three threads of evolution: nature, humankind, and technology. Each one possesses its own evolutionary logic. In this evolution, human technology is the new factor that ignited a new evolutionary engine.

The idea of a coevolution in which Technosapiens co-creates its own forms of life and interactions brings about a new framework of reference for human values: the environment, natural and technological, is not a scenery element but it is a stakeholder as a coevolutionary partner. Technoethics does not try directly to define the patterns of coevolution but it tries to define the emerging values. In this investigation, I will stress the importance of the ordinary. Coevolution, as evolution itself, achieves its extraordinary feats by the humble means of micro-events: a slightly improbable gene mutation here, a slightly innovative domestic use of an appliance there.

2.1 The Naturality of Technology: Human Nature

Philosophical Anthropology and Moral Anthropology

Philosophical attempts to characterize the anthropology of the technological human being are not new (Gehlen 1957). From the literature in the anthropology of modernity emerges the idea of a "techno-human condition" (Allenby and Sarewitz 2011). I will argue that Technosapiens' condition deserves a better fate than ideological exaltation or ideological abhorrence. I gladly concede that the naturality of technology and its *human* naturality, more precisely, are not perceptible when facing a nuclear plant or in a high-speed train. The naturality of technology, commanding a new technoethical approach, comes into sight on the other side of the electrical wire: at home, where nuclear energy lights a bulb under which someone is reading a book – the naturality and the human value of reading a book can be contested but they belong to a common-sense anthropology that I do not intend to put into question here. The naturality of technology is apparent again inside the high-speed train where a family is planning its summer vacation. These two examples might raise suspicion because they idealize the human use of technology but they intentionally point to value-laden instances: reading a book rather than vegging out with stupid media, talking together rather than vegging out with earphones on. Table 1.2 ("Ordinary Technology Domains and Instances") gives a list of essentially natural human situations and behaviors in technology: sitting in a car, spending time with a computer or a phone, using a payment card or making sure to take it when leaving home, buying or eating some processed food, taking a shower, and so on. Technosapiens' everyday situations, behaviors, and environment are saturated with technological artifacts that function in perfect harmony with human nature most of the time and that only rarely trigger a feeling of estrangement. However, to say that something "has become natural" (cars, screens, supermarket food, and so on) is most of the time intended to mean a deprecation of technology. Technology appears to be imposed on the "human part" of our lives because the stress in this sentence is on "become." The underlying narrative then is about a fictional primitive virgin human nature and what is afterward externally imposed on it. A technophobic evaluation ensues. I suggest that in the expression "it has become natural" we shift the stress to "natural" and take it in its full meaning. "Become" is to be understood in the sense that it takes in the phrase "we have become human" – referring to the cultural evolution that we call humanization, in which artifacts are included from the beginning, not as perturbing externalities but as contributive opportunities. *Werde was du bist*, "become what you are," said the German Romantic philosophers, meaning "you must become what you are" and expressing the fate of the human being.

From the Naturality of Technology to the Primacy of Ordinary Technology

The distinction between science and technology (Section 1.1) allows a robust type of naturality for technology. A first paradigmatic case is fire. Wood fires, used by human groups since an extremely remote period, have never been the implementation of a science or a theory of fire, in the modern sense of "science" and "theory." Fire was an efficient know-how supported by magic beliefs or the equivalent. What we can today consider as a theory of fire is hard science and it has something "extraordinary" about it, quantum theory and turbulence theory for example. Even the basic oxidation-reduction in combustion is an extraordinarily abstract and representational discourse. But fires have always been ordinary technology, on a totally different plane of evolution. I insist on "totally" different: a specialist in combustion physics will not have the slightest advantage over a boy scout in a competition of fire lighting and maintaining.

A "cognitive myth" prevents us from understanding the autonomy of natural ordinary technology. This modern myth asserts that every human capacity is in its essence a cognitive capacity. Then every human performance would be the application of a cognitive capacity. For every action, the human agent would have first to understand how it works (ideally through a scientific theory). This assumption clearly does not apply to fire, nor to bicycle riding, nor to the vast category of ordinary know-how skills that rely on no representational know-that at all. The emblematic technology of fire simply highlights how basic technology has always been to humanity. Technoethics tries to capture the pragmatic fact that human evolution and even humanization depended on wood fires more than thermodynamics, or to give a modern equivalent: Technosapiens' life depends on cargo ships and their containers (full of ordinary things) not on Archimedes' principle, the theory of their floating. A realistic technoethics needs to be attentive to the humble material civilization and to release it from the blurring influence of the "cognitive myth."

The second paradigmatic case for the naturality of technique in humans is not usually envisioned as a technique, but it is one: *language* is a physical technique, specific to the human body. It is not an artifact in the sense of a human-made object but it is an artificial performance. It is then a symbolic artifact, which is nonetheless material because sounds (or letters) are physical entities. There are good reasons to think that language is the disruptive human technique in evolution. I say "technique" here because I will use the term "technology" when another object than the human body is implied. Language appears thus as the paradigmatic case of "technique" similar to walking on our two legs, another remarkable case of human body technique which is perceived as natural. We apparently forget the efforts that were necessary to learn language and walking. This oblivion is evidence for the accepted naturality of these techniques, for their integration into human normality. It is no coincidence that both are ordinary activities.

Something unique about human nature resides in this capacity to integrate complex acquired techniques as ordinary, transparent, and pervasive functions of life. Something unique again occurs in historical evolution when spoken language, a body technique, undergoes a revolution consisting in its material transfer into the technology of writing, a technology made of material artifacts and devices. Then language undergoes another revolution consisting in the production and diffusion of printed language and, recently, a third revolution consisting in electronics and informational machines that facilitate human communication with apparently unlimited power.

Technology is as essentially natural to humanity as language is. It is a differentiating evolutionary factor as language has been and still is. Yet, the common representation of the major inflection points in human modern history emphasizes the *artificiality* of technoscientific discoveries, throughout the Industrial Revolutions: the First (eighteenth and nineteenth century, industry) and the Second (twentieth century, digitalization). This interpretation of recent evolution is in conflict with my stance on the primacy of ordinary technology. It is commonly thought that the historical disruptions were due to extraordinary scientific and designer-end accomplishments, not to natural ordinary use of technology. My rejoinder is once again the change of focus in ordinary technoethics. Even if technological revolutions are embodied by huge artificial systems, extraordinary by their size, their power, and their innovation, the revolutionary aspect is not an isolated technological parameter. The revolutionary aspect is not justified by the technological breakthrough in itself, an event in laboratories and factories, but by the transformation of ordinary life: now people travel by car, now people do not die from tuberculosis, now people can "Skype" each other from anywhere. This is the revolution; this is evolution, coevolution in fact. The everyday life of ordinary people is impacted by what constitutes a real evolutionary inflection point or revolution. The existence of a super-calculator or a super-robot that would be used by a few institutions (public or private) would perhaps not be seen as a technological revolution because of its lack in naturality and "ordinarity" – except by the media perhaps, but the hype would stop just after the advertisement campaign is no longer funded by the concerned corporations or government agencies. The Moon "conquest" in the 1960s, for instance, was perceived at that time as a major technological evolution or revolution because it was a media event, pushed into the life of ordinary people by the media. There were expectations of ordinary space travels and weekends on the moon that did not come true, simply because they never did become natural and ordinary.

The Human Natural Technosphere

We do not live in a science-fiction environment; we live in a natural technosphere. We dwell naturally in technology; this is the philosophical foundation for a constructive contemporary technoethics. Merleau-Ponty delineated the co-constitution between subject and lifeworld in many ordinary life examples. This tradition inspired Ihde's and the "post-phenomenological" school in philosophy of technology. The dimension of the ordinary is remarkable in these analyses of the interaction between humankind and its lifeworld. From wood fire and chipped stone-tools to the hot shower and vaccines, Technosapiens deploys an existential sphere that is inseparably human and technological: the technosphere.

Ihde (1990) provides ample arguments and illustrations in favor of the naturality of technology for human beings. The most striking evidence may be the presence of technological artifacts in prehistorical human tombs – weapons and jewels, but also ordinary life tools for eating or body care. This presence obviously does not withdraw anything from the humanity of the related culture. It works instead as the irrefutable argument for the humanity of a culture, which is always a culture that includes artifacts in the identity of a human person (Ihde 1990: 19). Let us test our ethical insights by the phenomenological method of "variation." Being buried with one's sword and war saddle raises no problem and the whole equipment will end after some centuries in a museum as an admirable testimony of culture – without much regard to what war exactly consists in. A virtuoso may ask to be buried with his/her violin; it should be possible but some would find it disturbing – let us suppose the violin is not a unique piece that would be lost for other players. Can a suburban youngster be buried with the motorcycle that caused his/her death? Why not? There is a cult book for techies that confers dignity to motorcycle engines and develops a sort of Techno-Buddhism – arguably not the queerest religious creed on earth (Pirsig 1974). A London City trader and a lot of us could be accompanied into the grave, as they are in very private moments, by their smartphone. I imagine how conservative and technophobic pundits would comment on this apocalyptic collapse of civilization and values. This variation series shows that the inclusion of artifacts in the sphere of human life, in a person's symbolic circle of integrity is an ongoing negotiation between *technē* and *logos* exactly in the sense meant by Hottois (1984). This ongoing negotiation is ordinary moral coevolution.

The ontological transparency of the technosphere is functional and contemporary technology tends to amplify it: I need to hear and to see the person I am talking to and not the telephone or computer screen, we need to catch meanings and to have a direct grip on our projects and objects without perceiving the intermediary instruments for these activities. This "intentional" transparency is similar to the transparency of the body in natural motion, where intentionalities (to grab a glass) are fulfilled without any perception of the intermediary body parts (arms and fingers). My intention to wash my motorcycle projects itself through the transparent means, low-tech (dust-cloth) or high-tech (high pressure washer). We live transparently in our technosphere exactly as the horse-culture peoples lived naturally in their own technosphere.

Human Flourishing in the Technosphere

As humans dwell naturally in the technosphere, they can flourish in the technosphere where they live. This human flourishing summons a new arrangement of Aristotelian ethical notions: the good for humans, technique, nature, the ends. I do not argue that human flourishing can be found in TV watching or Web surfing *by themselves*. I suggest that we go beyond the "by themselves" restrictive clause because in the technosphere these activities are always connected to a rich network of human and cultural engagements. With this background of communication and material facilitation, technology provides the initial and necessary conditions for human flourishing. Meeting basic needs, such as food and shelter, and providing comfort and reassurance on the level of material needs are affordances that do not "make" the human. Instead, they set the stage for making the human. This reassurance about the basic needs of life sets the conditions that allow human existence to yearn for more in fulfillment and flourishing. As long as we persist in considering technology, its facilities and devices, as external to human nature and human flourishing, we misunderstand the new existential platform for humanization in the present phase of human civilization.

The Worst Technoethical Paradigm: Natural vs. Artificial

The idea of a naturality of technology for humans leads to two critical conclusions, one against the "natural vs. artificial" ideology and the other against transhumanism.

The natural/artificial distinction is one of the most confusing ideas in common opinion and discourse about modernity. It can be used on purpose for manipulation but most of the time it is innocently referred to. Factually as well as philosophically, there is no borderline between the natural and the artificial. Our environment is a continuum of nature and technology. Ironically, two symmetrical sayings can be justified: "everything is natural" and "everything is artificial." The fields and cows in the landscape are artificial entities. In a car or a frozen hamburger, every component is natural in the end. These opposite slogans constitute an inconsistent semantic and a defective valuation method: if bread and wine, cow or dog, and the vegetables and fruits that we eat are all created by the technological agency of humans operating on a global raw material (nature, including living entities), then designating only certain entities as "natural" in a morally privileged realm appears as pure ideology and prejudice. This nonsense can be used sarcastically to rule out some deviant artifact in informal conversation ("Yuk! This ice-cream has an artificial taste"), but not as a serious value reference (to the "natural ice-cream" of old?). An interesting distinction is nevertheless maintained between nature and wilderness (no ice-cream at all in wilderness). In some cases, the anti-artificial semantic of "nature" refers to wilderness but this "deep" valuation reference is rare. Even hunting – the context in which Aldo Leopold celebrated the value of wilderness – is a technological affair, involving at least a primitive bow, which is as technological (philosophically) as a sophisticated rifle with night vision and a laser-targeting device.

My point is that the opposition natural/artificial never supplies an ethically acceptable ground of valuation. On the contrary, this distinction is a typical scapegoat: real issues (those that we cannot or do not want to address) are masked by attributing their evil to "artificial" technology. Technological scapegoats are perhaps easier to criticize, for instance when the Internet is made responsible for sex abuse or when TV is made responsible for violence.

The real scale of the problem appears in the informal and diffuse feeling that diseases are due to the abuse of technology (some are but not by essence and not by being an ethical sin) and that simple natural life would preserve health (it may, for those that never meet in their life a "natural" cause of death). We do not like the idea that a lot of fatal diseases are natural, whereas the drugs or procedures to treat them are artificial. We are more comfortable with the quasi-religious creed that diseases are the effects of our technology (sin) and that the curative and preventive treatments for them (salvation) must be natural. On the one hand we are suffocated by the comfort and facilities of technology, including the massive medicalization of any life inconvenience, and on the other hand we are flaunting the values of natural as opposed to artificial. In this context, the debate about GMO (Genetically Modified Organisms) in agriculture and food is typical: it is impeded by an ideological framework, the natural/artificial distinction, where genetics is taken as an industrial technoscience, artificial and therefore bad.

As is often the case for technoethical predicaments, the common flawed view coincides with technophobic doctrines. Therefore, without consciously assuming a technophobic moral anthropology, a lot of lay people contribute to spreading the vision of Martin Heidegger, Lewis Mumford or Jacques Ellul in a softened version that can be taken as common sense. The extremism and the ideological ambiguity of these doctrines should be kept in mind. For Heidegger, the initial Fall is ontological. Technology is an ontological catastrophe and alas the factual style of human existence. Ellul's background was more openly religious. His degenerative theory of human technology reactivates the grand narrative of sin. Even in Mumford, the sociopolitical analysis recounts the "myth of the machine" with technology featuring as the anti-Christ, expressing an anti-human hubris and imposing a self-destructive destiny. Directly opposed to this technophobic anthropology, my coevolution hypothesis and its moral pluralism consider the still possible construction of a dignified human person and a valuable technosphere.

The Second Worst Technoethical Paradigm: Transhumanism

Transhumanism inverts the opposition that I criticized in technophobia: value is on the side of the artificial and disparagement is on the side of nature and human nature. I will argue that transhumanism can be seen as an equally ideological reaction to the current technophobic anthropology. A philosophically stabilized model of transhumanism is circulating in the academy (Bostrom 2005; Savulescu and Bostrom 2008). It offers an explicit account of the moral issues involved (Bostrom and Roache 2007). Regardless of the media hype and self-promotion campaigns, the ethical questions raised by transhumanism can be sorted into problems and non-problems from a technoethical approach. An anti-nature ideology is applied to the human. I consider this approach to be the second worst technoethical model after the technophobic valorization of the "natural." This time technophilia is concerned, not technophobia, but the characteristics of ideology are exactly as conspicuous.

Transhumanism considers technology from the inside of an instrumental paradigm which turns out to be a compulsive optimization process. The project is to surpass and suppress the limits and flaws of human nature. Meanwhile, the perimeter of these limits and flaws remains indefinite as no substantial account of human nature is provided by transhumanist arguments. In the end, the idea is that human nature, as a "given," is in itself the limit and the flaw. Technological optimization is a value in itself and it evidently applies to the optimizing agent itself. As depicted in Table 1.3 ("The Three Kinds of Power"), transhumanism represents a sort of abomination: it takes the first kind of power, power over things, to apply it to the third case, power over oneself, the case where wisdom should prevail. This form of power over oneself is therefore maximally remote from wisdom. It is a pure case of technological command-and-control. The ethical problem is not directly the fact that the human person, mind and body, is treated as an object. This is acceptable in the context of health management, for instance, and more largely in any context of self-care: the instrumentality of technology is inserted into a network of conscious and lucid ends. The ethical problem lies in the type of power, in the one-dimensional command-and-control project, more than in its token application to the human.

The technological enhancement of the human self is a typical issue for technoethics. It can be addressed in its *ordinary* form. These ordinary instances are the cases where a choice is made between a technological fix and a self-care action. A significant case is offered by sleeping disorders, lack of natural sleep, to take the simplest case. What does it mean to "improve" one's sleep? In industrialized countries it is commonly taken as an optimization need, calling for a technological fix for which a pill is available. Fortunately, but after decades of erratic and dangerous over-medication (hypnotics in particular), systemic approaches are spreading. Some of them involve not only the "whole body" but also the whole person in a self-care project and a reform that extends to all aspects of existence.

The hubris of power in transhumanism was perceptible in the industrial version of its program (Roco and Bainbridge 2002), which was clearly business oriented and was more precisely a military-industrial program destined to maintain imperial power in the post-industrial era. The classic version of transhumanism tends to call its opponents "bioconservatives" (Bostrom and Roache 2007: 3) because its own power hubris considers ends like "mood and personality enhancement" in a continuity with medical curative technologies within an all encompassing movement called "progress," technological progress. This power hubris and this lack of reflexivity concerning instrumental rationality are captured in the expression "enhancement," taken as an end and the universal end – after its reduction to a simplified meaning predetermined by technological optimization in quantitative terms.

A different view of the cyborg, the future human merging with technology, must be evoked. There can be a naturality of the cyborg, not as a figure of the extraordinary, but in *ordinary* cyborgs: grandparents with a pacemaker or any cardiac implanted device, people with prosthetic limbs and organs that more and more merge with the "natural body." Clark (2003) extends the natural cyborg possibility and reality to the territory where the ethical becomes delicate: the mind. The philosophical anthropology that he defends is based on the naturality of technology for the human, as a characteristic of human capacities and activities. We use paper and pen to do math, which connects with our minds. "What is special about human brains, and what best explains the distinctive features of human intelligence, is precisely their ability to enter into deep and complex relationships with non-biological constructs, props and aids" (Clark 2003: 5).

The point that interests me the most here is the coevolution process where nature, human, and technology do not collide as "essences" – as if the purity of each were an intrinsic value. This prepares the notion of a self whose self-construction is not ideally derived from an essence but is self-realized and for this reason precisely bears an intrinsic value.

2.2 Technology as the Third Insider

New Stakeholders in Ethics: Nature, Technology

Technology is not an intruding outsider into the human lifeworld but it is an insider, a humble one, and one that has remained in the shadow for too long. The importance of material civilization was a discovery of historical research during the twentieth century, an astonishingly late discovery for our species, Technosapiens. Now that we are aware of the fact that civilization is not driven by political issues and artistic masterworks only (as it was pictured in history textbooks), we must beware of a compensating enthusiasm for technophilia. Civilization is not driven only by space travel, sport cars, and computers either. The present prominent visibility of technology is not a reason to promote technological progress as the sole agent in evolution as it appears to be in some high-tech business rituals launching new consumer products. Technology takes place in a coevolution scheme where nothing like one-dimensional determinism occurs.

In the dewesternized perspective of this book, deterministic visions of evolution are challenged by notions coming from Asian thought. The most important one is *interdependence* and its ethical form is *harmony* (Hershock 2014: 8–9; Hershock 2012). The challenge is to apprehend the constructive dimensions of coevolution within a new framework, centered on a global interdependence: binding together humans and nonhumans, living beings and artifacts, cultural and natural entities – infosphere, technosphere, ecosphere.

The problem with the humans is anthropocentrism. In our inherited vision, consciousness and its appendices (free will, intelligence, responsibility, and optionally an immortal soul) transcend every mundane reality. Consequently, history must be written on the transcending level of conscious human deeds and feats. This human transcendence entails continual progress. Any previous period tends to be seen as a dark or gray age. The coming periods will be brilliant with our illuminating productions, technological products, and the progressive human domestication of nature. Domination reigns, interdependence is not even perceived.

The origin of this narrow vision is our profound difficulty with the recognition of the other. It applies to the other human person, to other genders and ethnic groups, in general ethics. But it applies here to nonhuman entities, natural and technological. The philosophical discussion on the "intrinsic value" of natural entities in environmental ethics proved that "valuing" any nonhuman entity for itself, and not instrumentally, is possible but controversial. Now we need to value two global abstract "others," a second insider (nature) and a third insider (technology). The task is not only to value individual natural entities and technological artifacts, but to recognize these two global stakeholders and their specific teleology on the same level as humanity, i.e., the global abstract entity that we are.

Technology is mediating our connection to nature. This is just one of the three mediations involved in coevolution – humans mediate the relation between nature and technology; nature mediates our relation to technology. A triangle of mediation binds together the three stakeholders of coevolution. Mediation is taken here in the sense enriched by Verbeek (2011) and Van den Eede (2012), which focuses on the ethical dimension of this mediation. The first insight is the acknowledgment that we are not face to face with nature. Nature is no longer a block confronting us as it could have been perceived during the "conquest" phases (Nash 1973). A better vision of our present standing is given by the metaphor of immersion, no longer by confrontational images. Immersed in multiple ecospheres and technospheres, we are in fact immersed in the otherness of nature and in the otherness of technology. Indeed, we are creatures of the former and creators of the latter, but this a very one-dimensional link and there is so much more in our immersive coevolution than this instrumental production. In the end, what we have to figure out and assimilate is the agency of coevolution itself, resulting from the hybridization of three projects, three teleologies, three value frameworks.

Three Teleologies

In academic ethics, two value assets are now legitimate, human and nature. The priority between them is still under discussion but in my hypothesis this discussion (humanism vs. biocentrism) is not the relevant one, for two reasons: one of the three real protagonists is lacking (technology) and the issue should not be about priority but about coevolution. Technology becomes a serious candidate for a delegate moral value when it is understood as material culture. A moral consideration of technology becomes inevitable when technological systems are accepted as intelligent and even as sentient and emotional in some cases. This last stage of relevance for technoethics does not depend on the acceptance of an artificial intelligence that would qualify for any kind of Turing test (being indistinguishable from the human mind). Neither does it depend on Luciano Floridi's argument on the intrinsic value of information in itself. It is in the field of *ordinary technoethics* that this question can be addressed, beginning with the intelligent, emotionally and ethically rich relationship we have with our ICT (Information and Communications Technology) devices.

We have to consider three projects, three teleologies, three value frameworks: human, nature, technology. What matters is not a distinction of essences but of values and projects. It pertains to an ongoing evolution and not to the cataloguing of essences. In a pragmatic version of ethics: we are not reading essences, we are writing existences. The continuing existence of human culture was the only one that deserved consideration. Then came nature and its own evolution, the flourishing and extinctions of species, an autonomous teleology that for millennia was not impacted by human projects, or human means and ends, but that is now dramatically dependent on them. And finally technology sets in with its own evolution, developmental laws, and values. The evolution of technology can be understood through the analogies of natural evolution and human evolution.

Samuel Butler in *Erewhon* was amazed and afraid by the "natural evolution" of machines, which is faster than human evolution. This difference in the velocity of progress is a threat because the power of technology is going to overstep human power (Butler 1872). This simplified vision is typically confrontational. It is formulated in terms of power and supremacy. These two aspects made it easy to accept for the old set of values. Yet an important point is made by Butler and gives a good reason to be amazed, but not afraid. It is, for us post-Darwinians, the analogy itself between technological and natural evolution.

Gilbert Simondon suggested that an accomplished technology is always the result of an evolution with distinct evolutionary laws (Simondon 1958). For him this evolution is determined by an inner developmental logic and by the interaction with the technological environment. The two operating factors are (1) an inside evolutionary logic tending to completion and (2) the contingent occurrences of outside relational adaptations. The teleology of factor (1) is necessary to build coherent artifacts, coherent like organisms. This inner trend to coherence and functional integration in the evolution of complex artifacts constitutes for Simondon a specific trait in the evolution of technology.

Some philosophers consider that the evolution of technology, with its possible internal teleologies and with its human-driven teleology, has now taken over biological evolution (Hottois 1984: 133). We can imagine that we start growing ourselves wings (genetically) instead of building airplanes and in this direction the strongest version of techno-evolution taking over bio-evolution is to be found in transhumanism.

The diversity of our artifacts is as amazing as the diversity of natural entities. The amazing fact is innovation, which must not be seen only from the side of the self-indulging pride of creators but also from the side of the created, the objects, the artifacts. Basalla's study concludes that we have no satisfying theory of the actual factors in technological innovation and this is so mainly because these factors are too deeply intertwined with human and social factors (Basalla 1998: 134). But he maintains an evolutionary theory of technological innovation springing from an unfathomable source of variations upon which a selective process applies: Darwinian.

The enchantment of innovation is a powerful technophilic theme. The word "innovation" is obsessional in corporate parlance. But techno-evolution is at the same time a marvel and a threat of destructive disruption. A conversational model can be imagined (Nowotny et al. 2001: Chapter 4): social evolution and techno-evolution "talk" together, technology "responds" to a context and contexts respond in return. From a pragmatic point of view, this conversation does not happen through bureaucracy and expert symposiums, but it is rather driven by the ordinary use of artifacts and the ordinary micro-interactions in social and personal life.

Non-autonomous Technology

Langdon Winner's first important book has shown that the question of technology being "out of control" (or not) is a poor question (Winner 1977). Decades later, this still needs to be recalled in order to resist technophobia. If technology is a system in coevolution with two other global systems, nature and humankind, neither is technology instrumentalized by humankind nor is humankind instrumentalized by technology.

Inquiring into the origins of the "autonomous technology" tradition helps understand coevolution and its enemies. According to Ellul (1954)

and Mumford (1967), who are the major critiques of technological alienation, a conspiracy of the "System" against decent civilization must be
exposed. It does not really matter whether the final conspirator is business
("capitalism") or technology: the autonomy of this anti-human process
means conspiracy in itself because it deprives the political level of its steer-
ing power, in market democracies just as much as in authoritative states.
For ethical analysis, this is the classical problem of means and ends: technology (or business) has become an end in itself.

A latent fear of technological hubris may explain the reluctance to
accept technology as an evolutionary stream, which flows in the midst of
interdependence and which can be in harmony with human evolution and
with natural evolution. Would technology run amok if it achieves agency
or autonomy? The reason for this somber prognosis is a form of ontological intolerance: the prejudice against an entity that is neither human
nor natural. A computer being too intelligent and talking with a soft persuasive feminine voice in the beginning of a fiction episode has a very high
probability of ethical reversal before the end: from means to ends, from
instrument to subject, from being dominated to dominating. The narrative
fertility of this scenario does not prove it is factually or philosophically
sound. Retracing it back to its original inversion scheme highlights the
question of power. It can be put in its simplest terms: either we dominate
technology or technology dominates us. The same simple terms apply to
our relation with nature: dominate or be dominated. Hopefully, we can
imagine different relationships with technology and with nature (care,
harmony, respect, sustainable interdependence). This ethical innovation is
on par with the non-domination relationship that we need to build
between humans. What sense does it make to say to one's smartphone:
"Either I dominate you or you dominate me, we have to settle this!" Yet
for some of us the fact that we are the masters of our phones, computers
or cars could occasionally be doubted.

The power in question is power over things (technology, Table 1.3)
transposed into the terms of power over people (politics, Table 1.3). In the
simplified question of technology as power, the solution is simple: in its
present "alienated" and "alienating" state, technology is not in the right
hands. It is either in bad hands or in no hands at all. More than often, the
persons or groups making this point tend to consider their own hands as
the right hands, which amounts to saying that they want power. The question of technology is missed, again. This time it is missed as an ethical
question and the whole affair returns to business as usual in politics: competition for power. Understanding technology is another affair.

Examples of interactive coevolution with technology are everywhere
around us and the most ordinary of them are the most ethically revealing.
Let us examine railroads and their multifarious substructures. All the
implied entities, rails and trains but also water and coal distribution, constitute an exemplary technosphere. Its shape is not spherical. Geometri-
cally, it has the shape of a web stretched over the countryside. The railroad

technosphere is hybrid because the countryside and many natural entities belong to it: plains and mountains, deep valleys and rivers to cross. Bridges and all civil engineering artworks included in the railroad technosphere have profoundly modified the natural landscape in industrial countries. Human actors are included as well in the railroad technosphere: the engineers who design railroad artifacts and try to "sell" them to investors and politicians (Latour 1992), the passengers that sometimes feel they are a negligible quantity in the process, Irish and Chinese workers that built the railways in the American West and are forgotten – even if Thoreau wrote that their dead bodies are symbolized by the "sleepers" (crossties of the railways tracks), so that "we do not ride on the railroad, it rides upon us," humans (Thoreau [1854] 1985: 395–396). In the web of the railroad, natural and human entities are tightly interwoven. Their history and evolution are inseparable. The same picture can be given of the electrical network deployment history and virtually of every transportation or communication network.

This scheme and its technoethical consequences are all the more apparent in the case of the central artifact in our industrial society: the automobile. A true *car cult* characterizes our civilization. It alone can explain our industrial structure, our constant yearning for fossil fuels, and from there it can explain a large part of contemporary geopolitics, tensions, and conflicts. Car cult alone can explain the shape of our cities and landscapes, explain where we accept to dwell and how we spend a lot of our time and of our money, or why we need to fight every day against being overweight. What I want to emphasize in this disparate list of facts and habits is the hybrid nature of our existential sphere.

The Darwinian Logic of Coevolution

Coevolution involves a Darwinian process of a new kind. The question is about its moral acceptability. To address coevolution ethics we need to specify the precise level of the operating process in this evolution, which in my view is the ordinary life of the individual. Darwinian evolution in biology can be conceived as not happening on the level of species and millennia but on the level of genetic micro-changes and selective micro-events, all on the modest level of the individual organism or even the individual cell, in anonymous corners of life and at any time. My point is that the coevolution of technology, nature, and humankind happens on the same level. It is driven by the integral summation of innumerable micro-events and microactions, following a Darwinian model of emergence and not a mechanical model of domination (by a supernatural entity or by conspiracy).

Jon Elster gives some references of similar conceptions of technical change as evolutionary processes driven by micro-events (Elster 1983: Chapter 6). An even better expression of this view uses the term *exo-Darwinian* evolution, promoted by Michel Serres in philosophy. In his view, we have externalized our bodily functions into technological artifacts (Serres 2001). This idea takes in its literal meaning the classical metaphor presenting trains as super-legs or the recent metaphor presenting computers as brain annexes. These are not images but facts of coevolution, which is exo-Darwinian in the sense of exo-biological. Exo-Darwinism in this acceptation claims that a second engine of evolution has been ignited. This engine is not "natural" but human-made and it raises for us preoccupying questions of responsibility.

If the coevolution of nature, technology, and humans is now exo-Darwinian, does it mean that there is a pilot in coevolution and that we are this pilot? A simplistic technophilic answer is "yes." Politicians and advertisers frequently utter it. Technoethics works on a different agenda and puts forward a different model. Human nature in us, Technosapiens, implies the paradox of a natural artificial species (Section 2.1). Not only are we this species but more and more we are its creator through the coevolutionary function of technology. The relative privilege of humankind is now to understand itself as the acting leader in a system loaded with irreversible menaces, and menaces particularly ominous for the (temporarily) weaker entity in coevolution, which is nature. The bonds of interdependence are easy to identify. Technology is dependent on the existence of humans, obviously. But humans are dependent on the existence of nature without which they would not exist and prosper much longer than computers or cars would last without the service rendered by humans. The evolution of nature, now, is more impacted by the action of humankind and of technology than by its own internal Darwinian process.

In a sense, then, humans are the environment of technology. Nature is the environment of humans and of technology (material artifacts are made out of natural elements, mineral or organic in their origin). Technology is the environment of humans, particularly since the Neolithic period. Technology has now become the environment of nature on large surfaces of the planet where the major ground occupant is urban concrete or agribusiness soil. Nature is invited to fill the interspaces where its recreational or productive presence is required. Human communities are now hybrid communities (Latour 2005: 247) and these hybrid communities are eco-systems and techno-systems at the same time.

2.3 The Present Stage and the Potential for Coevolution

Once the principle of coevolution has been clarified the next question for technoethics is to assess the potentials of this coevolution between nature, humankind, and technology, not in itself and in the abstract, but in the present state of the world. From a pragmatic point of view, what I call here *potentials* refers to beings and facts that are neither to be feared (technophobia) nor to be passively expected (naive technophilia).

From Coevolution to Harmony

Orienting coevolution toward harmony is not exactly a matter of the human will. Coevolution is systemic and no linear causation can be used in it. The "Columbian Exchange," for instance, (the systemic exchange between the European and American continents) intertwines *ecological* imports (microbes, plants, animals), *technological* imports, and *cultural* imports. Important imports typically belong to the three categories at the same time (nature, technology, humankind): horse culture and Old World agricultural techniques.

The conceptual bridge that we need to build goes from coevolution and interdependence toward harmony in a systemic or symbiotic balance. Harmony is at the same time an ontological and a behavioral notion in Daoist (Wong 2011) or Buddhist thought (Hershock 2006). Harmony defines the form of interdependence that is optimal in itself and that we must try to realize in our lifeworld. The Dao offers here a better ontology of process and change than our current technoscientific ontology. In this view, what really exists and what produces every entity is change.

Harmony is one of the virtues for ordinary technoethics (see Table 1.4, "Wisdom") and it is also a teleological condition for the ethical endeavor itself. This last point is made by virtue ethicist Rosalind Hursthouse: "[...] the practice of ethical thought, as we know it, has to be based on the assumption that human beings, as species, are capable of harmony, both within themselves and with each other" (Hursthouse 1999: 265). My suggestion is to continue and to enlarge the scope in this way: "and with their environments, both natural and technological, and with all hybrid entities emerging from the coevolution of the human species and its environments." Harmony makes sense in all possible forms of interdependence, within humans as well as within the ecosphere and the technosphere. This can be the ethical project of modernity.

Harmony is inscribed in the past of evolution and coevolution. With all due respect for the confrontational logic that pertains to Darwinian competition and selection, we still need to concentrate on *collaborative* coevolution. An ecosystem is by its very nature a collaborative system. In moral anthropology, I fully accept Benkler's vision of a naturally collaborative (and not competitive) human agent, a vision buttressed by ordinary technologies (Benkler 2011). In the history of technology almost every recent artifact results from the collaboration and operational merging of varied engineering traditions and skills and the coalescence of varied subcomponents.

Presentology and not Futurology

The relevant research in technoethics is not about the future but about the present. The reappropriation of the present differs from indulging in technophilic futurology because the latter is a passive imaginary satisfaction

and the former is an active continued effort. In a sense, futuristic specula-
tion is not innocent. It can be considered a voluntary fictional distraction
from the ordinary, from the needs and from the resources of present
reality. To live in the present, to be fully aware of it and not intoxicated
by media representations constitutes the sound basis of an ethical reappropriation of the ordinary and thereby of the real potentials for coevolution. What happens in a real case of medical emergency, for instance, is
very different from the fictional representation of Emergency Rooms in
TV fictions. Most of the time, the availability of medical support and
treatment will vastly differ in real cases from what is displayed in magazines and media reports. Sticking to reality and refusing to live in the
sphere of representation requires a conscious orientation of the mind. As
Borgmann puts it, "holding on to reality" is a virtue that we need to
develop (Borgmann 1999). We do not really need to project ourselves in
the future as much as we need to project ourselves in the awareness of the
present and its potentials. If not, the possible cure for a disease may
remain a futuristic expectation, used by some to raise money and thrive
on largely fabricated extrapolations, while in the present the method
to prevent this precise disease are already available and accessible by
taking care of one's way of life and of one's environment – safe sex, sanitation, sane food, drinkable water. Political and technoscience agendas
are efficiently promoted by a "tyranny of prophecy" (Dublin 1992) which
openly appeals to religious aspirations for salvation and divinity (Noble
1997).

However, where is today the existing technological utopia? Fictions are
almost all the time dystopian, seeing the future as distorted by technology
and never again, it seems, as enchanted by technology. Actually, technological utopia exists and futurology prospers in advertisement, explicit or
implicit, which occupies a large part of the media. GAFA (Google, Apple,
Facebook, Amazon) and not Hollywood furnishes us with positive futurology. The short movies of futurological paradises to be found on the Web
will be corporation ads. The trailers of fiction movies will be scenarios of
catastrophes, with rare exceptions. Presentology is an effort to get rid of
this advertised shallow futurology.

The media commonly "sell" advertising fantasies about technological
gadgets simply by extrapolating the comfort benefits and the psychological
reassurance of ordinary life in technology. They validate by this extrapola-
tion the present trajectory of industrial and political institutions. Rejoice,
iPhone$(n+1)$ is going to arrive and by essence it will provide more comfort
and power than iPhone(n), the one we sold to you last year by the millions.
The consumer-citizen has nothing to do, because the future comes by itself.
Just wait and prepare to pay. Behind this "promise" of technology, a
message to the consumer-citizen is put forth: the industrial system and its
political structures must go on if you want this brave technological future
to happen. In representations of medical progress, the futurologic tension
is even greater than in representations of recreational technology. There is

no incurable condition, it seems; just pathologies for which the cure has not yet been invented by technoscience. More time and more money is all that is needed. Just wait and pay. And make sure that nothing perturbs progress. When aging or even death starts to figure among the pathologies that are waiting for a technological fix, the hubris of futurology reaches the absurd.

My point is that the illusions of futurologic projections stand in the way of the real potentials of technology. This is the case not only because of stupidification in consumer society and because of genius in the advertisement industry. The situation comes from identifiable and remediable flaws in our understanding of technology and coevolution.

The Promise of Technology

Borgmann's notion of the "promise" of technology (Borgmann 1984: Chapter 8) evokes both the ideological futurology that I want to avoid and the constructive potential of coevolution that I advocate. The satisfactions brought by our technological domination of nature command the script of Western civilization since the Enlightenment, says Borgmann. Fulfilling this promise is key not only to our vision of the world but also to our values system. This analysis can make the most of two semantic ambiguities in the term "promise." The first one is the particular meaning of "promise" in the advertizing industry ("the brand promise"): the gratifications that can reliably be expected from a product and that are the base of the trust relationship a brand asks for and advertizes for. The second semantic ambiguity is the pejorative connotation of "promise" in current use ("only promises" or "false promises"). The term is therefore a good one to capture technology as it is perceived today.

Nevertheless, a real potential is expressed in the form of this "promise" of progress and modernity. The promise still carries the energy of a liberation movement but it must be reinterpreted in the perspective of a coevolution aiming at harmony and flourishing, which implies leaving at bay and ultimately forgetting the old values of domination and confrontation.

The end of the industrial era (Roszak 1972; Bell 1973) marks the end of traditional hopes in the institutional social engineering that was called "politics" during this era, from the eighteenth to the twentieth century. A non-confrontational model of coevolution is better suited to the post-industrial era. The change of focus from production (and design) to consumption (and use) is central. In the realm of consumption the post-industrial era is commanded by an amazing phenomenon: affluence. Evolution has had no time for rewiring the human brain, which is running an operating system based on scarcity, and for inventing an operating system configured to manage affluence. Consequently, our adaptation to affluence is handled by culture, the software in our brain. The normal regime for economy and society is scarcity and the competition for rare resources. But the present state of coevolution has placed humankind in front of affluence. The result is a massive new challenge, the sustainability of affluence. Technology, progress, and modernity were a "promise." It is fulfilled in such a way that it turns into a problem.

Laws and Super-laws of Coevolution

Some laws of coevolution can be observed at the interface between human-kind and technology. A simple system of attraction and repulsion poles in the present evolution of artifacts suggests a list of attraction poles: mini-aturization, wearability, personal devices, informational devices. The repulsion poles would be: visible prosthetic device, material infrastructures, inert tools, collective (industrial) devices (Puech 2008: 74–78). Four law-like forms of evolution of artifacts can be discerned: a *miniaturization law* (smaller is better), a *dematerialization law* (digital is better), an *integration law* (one-does-all is better), a *simplification law* (simpler is better). Super-laws, abstracted from these synthetic laws, can make sense and be inspir-ing. Two super-laws can be derived: a super-law of *informationalization* (from mass and energy toward information) and a super-law of *personali-zation* (from collective devices to individual and embodied artifacts). Both of these tendencies are significant for the coevolution of the humankind/ technology interface because they converge toward a greater proximity between us and technological entities. In the present evolution, artifacts tend to be *proximal* entities. The informationalization super-law brings them closer to our mind, the personalization super-law brings them closer to our body.

A final synthetic super-law would then be the law of *proximality* – which invites ethical assessment. In the present stage of evolution, smart-phones and personal computers, and behind them the digital technosphere, constitute a *proximal* technosphere.

The Technosphere Revolution

The global disruptive innovation of today is the technosphere, a notion that synthesizes the essential new data of coevolution. Humankind and nature are immersed in a technosphere. This immersion is not only phys-ical because the technosphere is partially immaterial and digital (the info-sphere), so its pervasiveness is not only geographical and spatial. This is not a situation of domination or manipulation; the technosphere is not a Matrix as in the science-fiction movie. The technosphere is an amplifica-tion of the human sphere, not in the spatial dimension but in intensity and in density. The human-made environment is now deep and thick every-where. The infosphere is a determining part of this actual pervasive tech-nosphere. Technosphere then means the complete set of the technological environment, both its material and digital components. The spherical prop-erty symbolizes the fact that the whole process takes place on a planet, Earth. The sphere is also a symbol of the "environing" property of modern technology, its capacity to provide an environment for the two others (humankind and nature).

I will argue that the real nature and the most important features of the technosphere are due to its being largely an infosphere. An interesting leading thread through coevolution then goes from spoken language to written then printed language and finally to the electronic management of information. Every stage is a revolution and our technosphere/infosphere is the cumulative sum of these revolutions, each of them a cultural revolution inside the material revolution.

The potential of our time is concentrated in the present state of a digitalized technosphere, an opportunity that could not have even been imagined before. In his enthusiastic *Being digital*, Negroponte (1995) was celebrating potentials, opportunities to be realized, and not achievements. At any stage, digitalization is not an end in itself but a powerful means. For this reason, the present stage of coevolution requires an assessment in terms of ends for these means, in terms of projects, engagement, importance, and value. Technoethics works for the sake of this constructive evaluation. I agree with Dreyfus about the Internet: "The Internet is not just a new technological innovation; it is a new type of technological innovation; one that brings out the very essence of technology" (Dreyfus 2001: 1). His statement is even more valid with this precision: "If the essence of technology is to make everything easily accessible and optimizable, then the Internet is the perfect technological device" (Dreyfus 2001: 1–2). But beyond facilitation, access, and optimization, all of them valuable directions for instrumental rationality, the technosphere/infosphere prompts an open program of disruptive innovation and not only the incremental improvement of what we have. It is well known that software is lagging behind hardware: it is difficult to imagine how to use the full power of our technology in a really creative way. In a sense, the technosphere is the hardware for possible lives that are still to be invented.

Ubiquitous Mediation

When discussing "ambient intelligence" (Tavani 2004: 362–369) or more generally the pervasive presence of "bits" in the world of "atoms" (Negroponte 1995; Jurgenson 2012), recent philosophy of technology is in fact exploring technological *mediation*.

Disruptive potentials are available at both ends of the infosphere: in the proximal technosphere, with wearable computing, ambient or ubiquitous computing, and in the distal technosphere of networks. Manuel Castells has extensively investigated the combined changes in the self and in society in a series of books published just before the turn of the millennium (Castells 1996, 1997, 1998). The following decades in human history confirmed his analysis. Technoethics develops this framework with a focus on the notion of *mediation*, taken in its existential and ethical meaning. Van den Eede (2012) pursues an original course of ideas that found in

MacLuhan the inspiration to say: every technology is a media – the reverse fact is obvious: every media is a technology.

Universal mediation by technology (Verbeek 2011) must be systematically interpreted as a mediation in its full right and this leads to the idea of the technosphere itself as an ubiquitous media or ambient mediation. One material aspect of this situation is the "Internet of Things" that we are starting to implement. The most relevant form of the phenomenon lies beyond cables and electromagnetic connections: it is an existential situation. Human existence today is lived in an environment of networks that mediate every action and every relation. Formulating the questions in terms of ubiquitous mediation is an opening for a post-ideological interpretation of human agency in the ecosphere and in the technosphere.

Ecological Consciousness

The accounts we give of our actions (history) and of our invention of artifacts (history of technology) have the tendency to entirely forget the dimension of nature. The eighteenth- and nineteenth-century industrial revolutions in England, for instance, were observed in two dimensions: the technological revolution and its human consequences. Consequences on nature could be mentioned but as an appendix and as far as they impact humankind. The idea is not only to call for a history of nature in eighteenth- and nineteenth-century England, but also for an integrated history. The new element in our worldview is this ecological consciousness, the simple fact that we are beginning to "see" nature, to take her into account and as a character in coevolution rather than an externality. Our vision of technology in the environment has changed (Tiles in Olsen et al. 2009: Chapter 42).

The emblematic "Gaia hypothesis" by James Lovelock makes nature the main character (Lovelock 1979). The ecosphere or the biosphere can be taken as a living organism with its own evolution and flourishing. This assumption brings along a values system that is identical to the values carried by any organism in environmental ethics. The difference is that now humankind is just one component of this organism. It then deserves the status of auxiliary or instrumental, deprived of original intrinsic value. We should not play musical chairs with value carriers (souls, humans, Gaia). But we can take Gaia (nature) as an essential partner in the complex game of a common evolution, as a partner with its own agenda and set of values. Technoethics adds a third dimension to the framework: technology.

Two of the coevolving entities, humankind and technology, are so detrimental to the third one, nature, that the present regime of coevolution is not sustainable. This predicament is known to us humans since at least the 1970s. The factual reality of an unsustainable regime of coevolution is the darkest potential of its present stage. But ironically this new consciousness arose in an intellectual process containing its own positive potentials.

The brightest of these new perspectives lies in the very notion of ecosystem. The internal coevolution and symbiotic workings of ecosystems offer a paradigm and better: a real case of the harmony in coevolution that could make sense on the global level. The ecosystem is a value carrier and a value model.

This can be the challenge for the technological age: being ecosystemic and therefore sustainable. It is nothing but a version of what Borgmann has identified as "the challenge of nature": "the realization that nature in its wildness attains new and positive significance within the technological setting" (Borgmann 1984: 182). The ecosystemic values experienced through the contact with "pristine" nature and with ordinary nature everywhere (from the oceans to our own body) has a cognitive significance as well as an ethical significance. Emerson and Thoreau were the first in Western culture to detect that "nature's lessons" have this double dimension and that the foundation of ethics in modernity must start from this experience.

Important technoethical tenets are directly transposable from environmental ethics. From Taylor's research on "respect for nature" we can learn how to extend ethical significance to nonhumans, moral subjects that are not necessarily moral agents (Taylor 1986: 17). Taylor's effort to enlarge ethics typically leads to lists of virtues: conscientiousness, integrity, patience, courage, temperance or self-control, disinterestedness, perseverance, steadfastness-in-duty (Taylor 1986: Section 4.3) and to a list of "moral concerns," including benevolence, compassion, sympathy, and caring (Taylor 1986: 205). In the end, his "ethical ideal of harmony between human civilization and nature" (Taylor 1986: Section 6.4) is ready for a technoethical enlargement.

Hybrid Ethics (Ordinary)

Our post-modern ontology situates us in a universe of hybrid objects. We need to include in this cross-hybridization all the three domains – nature, humankind, and technology. What could be the hybrid ethics subsequent to this ontology?

Let us try to see ourselves as partly "artificial selves" or as selves merging with technology The human's technological ("artificial") dimension is now circumscribed by bioethics and medical ethics. Is the emergence of *artificial selves* the last step in this process? I will argue it is not so and I suggest that the emergence of *nonhuman agents* is a more pregnant issue. Nonhuman agents do not need to be "selves" in order to be agents for which accountability and responsibility make real sense. The ethics of their relation with humans is the problem for technoethics as long as technoethics is thought by humans. In the perspective of this book, bridging the gap between human selves and artificial agents belongs to the ordinary confrontation with "smart" artifacts.

The analogy with animal ethics can help to make this point. The ontological questions about mind and mental states in nonhuman animals are not about essences but rather about practical interactions. A full range of emotions and attachments is experienced and these data constitute the relevant basis for an ethical approach. Human/animal interaction yields without any doubt a rich set of values. Mary Midgley's model of a mixed moral community (Midgley 1984), reminiscent of an idea attributable to Naess (1979), invites including artifacts in the community in order to finally envisage a larger idea of mixed communities (local and ordinary) made of human, natural, and technological entities. These mixed communities would be the best realization attainable from today's potentials in coevolution. The source of ethical values and moral significance in the functioning of these mixed communities would not depend on the ontological status of their members, in particular their being "selves" or not. We are selves; we are arguably in the process of evolving for ourselves new forms of being a self, but the intrinsic value of an entity is not proportional to its resemblance to the human self. When "having a value" is equated with "resembling us," a dangerous confrontational values system reigns.

On the same line of argument, the ethical question is not about the exceptional ontological status of "artificial selves" appearing here or there in the technosphere. What matters are the ethical bonds created by the coevolutionary emergence and preponderance of these mixed moral communities where humans and artifacts, and also natural entities, shape together and share a mixed lifeworld.

2.4 Human Threads in the Coevolution Braid

The Human Agent: We Are It

Technoethics is thought by and for human agents. The privilege of the human agent is thus obvious even if limited: we are it. This privilege is a fact and not a value. It is a perspective advantage that confers a sort of "wise anthropocentrism" upon us in technoethics. Biocentrism or technocentrism would be unwise. Anthropocentrism was also unwise because we are one of the three in coevolution. It would be arrogance again, after ignoring the two others, to speak in their name. Technoethics is definitely a human discipline; it does not pretend to talk in the name of others, neither trees nor computers.

In this modest approach, the human agent has no particular "duty" to survive (Derringh in Hershock et al. 2003). Awareness of humankind's role in recent coevolution could even suggest a duty not to survive for the sake of the ecosphere. In a constructive scenario, rather than having a duty to survive, the human agent can seize an opportunity to flourish and to dwell in harmony on the planet. In this project, the issues ahead are not technical but ethical.

Another privilege of humankind in coevolution is the disruptive pace of its own evolution. Our agency in the last two millennia oversteps by far the rhythm of change on the planet. Bio-evolution is slow; variation and selection are even necessarily slow because they need long-term processes for the statistical engine in them to run efficiently (Dennett 1995). Techno-evolution has been fast since the Neolithic period, and it is rapid in our twenty-first century. Human evolution stands in between. The present technosphere educates the human into perpetual innovation and learning. Nevertheless, the important thing is not who or what is faster but who or what sets the pace for coevolution. In the Anthropocene, we are shaping the planet, for better or for worse. The question is not about the right that we might or might not have to do so because there is no transcendent dispenser of such rights. We have used this right and now the question is formulated in Hans Jonas' principle of responsibility (Jonas 1979): we do it and we know it, therefore we are in charge of it; we must be accountable for it. In other words, we are responsible for the planet even if we do not want to be.

Affirming Human Values

Human values are interpretations but interpretations shape the world through human projects. The Asian notion of *karma* is welcome to temper this all too Western voluntarism. The Western vision of historical engineering and management must be modified to fit with coevolution and the moral interdependence expressed by the notion of karma (Harvey 2000: 8–9). According to Hershock, the modern predicament can be reconsidered from a karmic point of view:

> [...] even if the *facts* of our present situation are relatively resistant to immediate alteration, the *meaning* of this situation is always open to negotiation. In karmic terms, responsibility always entails the possibility of meaningful response. While the particular situations in which we find ourselves are invariably the conditioned *outcome* of prior intentions and continuing values, these same situations are also sites of *opportunity* for revising presently obtaining patterns of value and intention.
> (Hershock 2006: 106–107)

Human threads in the coevolution braid are visible. Do they follow an acceptable value project? The records of our history, from the shorter range to the most extensive long range, display no evidence of an "intelligent design" of human values enacted by human agents except for the value of domination. Reconceiving human agency in the context of coevolution does not imply isolating humankind as the sole carrier of agency and projects in the whole universe. Coevolution means neither more nor less human agency but rather a different kind of agency: projects that take into account the global environment, natural and technological with its intrinsic values and its hybrid and nonhuman agencies.

The achievement of human values at the scale of global coevolution cannot be expected from international institutions and politicians: their inertia on that matter is what led us into the present civilization crisis. I argue that the virtues of ordinary technoethics constitute a new program of action to start with. The cardinal virtue of wisdom detailed in this book characterizes the approach to an alternative human agency. Its six applicative practices (Table 1.4) make sense for coevolution: *awareness* in order to shift our vision toward global coevolution; *benevolence* in order to take into account the intrinsic values of nature and of artifacts; *autonomy* in order to avoid being dominated (by technology mainly); *humility* in order to avoid dominating; *harmony* in order to merge into coevolution; *courage* in order to remain an agent and a value enacting agent.

Borgmann quotes Churchill's principle: "We shape our buildings, and afterwards our buildings shape us" (Borgmann 2006: 175). He adapts this principle through several ethical variations. I simply add this technoethical version: we shape our ordinary environment, and then our ordinary environment shapes us. Environment here encompasses natural as well as artificial environments, including cultural and digital environments.

Human agency can operate on three levels: (1) globally engineering the world, (2) designing one's local environment, and (3) enhancing the self. The harmony between these levels is as important as the interdependence between them. I will insist on the correlation between (2) and (3) according to Churchill's principle. I will then suggest that the priority of (1) is questionable and that we must not let it impose a top-down governance on our life. On the contrary bottom-up methods encourage the reverse order: from the dynamic interaction between (2) and (3) toward the emergence of new patterns for (1). Thus, the affirmation of human values inside the harmony of coevolution is not an engineering design (accepted off-plan) but can be compared to the continuing emergent transformation of an ecosystem (bottom-up from micro-changes) and to the constant evolution of the Internet (bottom-up from micro-uses).

Progress and Loss

"This history of American transportation shows that a wealthy nation can make decisions that impoverish rather than enhance its choices," states Nye (2006: 220). Do we need to reconsider what we formerly took as progress with unshakeable faith? Nye tellingly suggests that we can change "how" technology matters for us. Progress might not be the adequate notion for this. Neither the positivist version of progress nor the postmodern eradication of the notion is sustainable in the framework of technoethics.

Nevertheless, the notion of progress remains a pragmatic mobilizer for action behind the dogma (Nisbet 1980: 8–9). To restate progress as a question and possibly as a value we must surmount the dogma of progress. Its origin lies in the program of technoscience. Its triumph in Western ideology is summarized in the expression "command-and-control." Even before the classical formulation of the doctrine of progress by Condorcet in the

1790s, another francophone philosopher gave a more complex image of human evolution, Jean-Jacques Rousseau. Rousseau's well-known critique of progress in two early essays (*Discours sur les sciences et les arts*, 1750; *Discours sur l'origine et les fondements de l'inégalité parmi les hommes*, 1755) is unambiguously technophobic. However, the fine grain of its analysis appears in his more mature and thorough books (*L'Émile*, 1762 and *La Nouvelle Héloïse*, 1762). Rousseau depicts the personal destiny of the human person and the collective destiny of societies through a specific notion of change. Change is interpreted as a disruptive loss, irreparable, but after which life goes on. At first sight, this is one of the inconsistencies of the mocked Jean-Jacques. He cries out chapter after chapter "everything was lost at this instant!" and after some pages of lamentation, life starts again in the biography of persons (himself at first) and the history of societies. When Rousseau is read as a profound and original thinker, surpassing the Enlightenment/Romanticism dilemma, this formulation can be read as an innovative vision of the irreversible loss in human evolution. In Rousseau's concept of irreversible loss what matters is the deep sense of mourning and the resilient reaction. The contradiction is only apparent: acknowledging the loss, the fact that what is lost is irrecoverable and irreparable, but at the same time having the capacity to surmount this fact, to live with it without denying it, to go further. The paradigmatic application of this pattern is the relation between humankind and nature. It was central for Rousseau, both on the individual and on the collective level, as it still is for us: our original immersion in nature is lost and cannot be retrieved, but now our task is to preserve the possibility of authenticity while the natural is lost.

In the coevolution between humankind, nature, and technology, losses are to be acknowledged and surmounted. Our effort can be to reinvent a regime of harmony between the three rather than prevent any loss at any price. A new technology can cause an irremediable loss but life goes on and the challenge is to re-harmonize the new context. Frozen food and microwaves, for example, changed the way cooking is conceived. Large parts of tradition in the supply chain and processes of family cooking are gone. But in a kitchen equipped with contemporary devices a re-harmonization is possible, through which healthy and excellent home-made food can be preserved frozen for instance. Another use of the same devices would cause imbalance: junk food habits, life hygiene degradation, ecological damage.

How Better? Flourishing and Dwelling

"Degraded environments are necessarily correlated with degraded patterns of consciousness" (Hershock 2006: 16), but the opposite is also true: restored patterns of consciousness harmonize restored environments, natural and technological. What emerges from the reappropriation of a more authentic relation to (ordinary) realities in nature and technology is simply human flourishing. The intrinsic value of flourishing is not limited to humans. It hints at a common logic of values in coevolution, for entities in nature, technology, and humankind. The human agency to harmonize flourishing contexts can be seen as a new stage in the flourishing opportunities of the universe.

Flourishing is certainly the best notion for reinterpreting and perhaps replacing the idea of progress. The value of progress implies a global perspective in which every moment or entity is just a means, an episode, a point on the graph line, whereas flourishing always offers a place to stand, an entity to be. Flourishing can be an end in itself or an intrinsic value. Progress dispensed global externalities, flourishing settles local values.

In a 1992 text about Tvergastein, his home in the wild, Arne Naess pleaded for the restoration of a "sense of place" in modernity (Naess 2010: 45–64). Dwelling means attachment bonds, memories, belonging, and a sense of flourishing in connection with an environment. The fundamental harmony with contexts is expressed in the human mode of *dwelling*. Coevolution conveys a broader image of dwelling: we dwell in nature and in technology; our agency in dwelling constantly reconfigures the natural and technological environment. We dwell naturally in technology and we dwell technologically in nature: this constitutes the specificity of human dwelling.

The central notion of dwelling can be understood from Heidegger's cryptic but essential references to the poet Friedrich Hölderlin and from Heidegger's conference "Bauen wohnen denken" (1951). The links between building, dwelling, and thinking underlie the human existential structure and characterize human activity. They must be explored in all possible combinations and in a more technophilic mood than Heidegger's. For this purpose, inspiration can be found in Ivan Illich's own characterization of the art of dwelling as a typical human art, an art that got lost and needs to be restored. Illich's vision was put forth in a 1984 address to the Royal Institute of British Architects entitled "Dwelling" (Illich 1992).

Modern cities and buildings are spatial commodities that are paid for and consumed. The challenge here again is to invent the possible harmony in this context, which means to implement technology, ordinary technology specifically, as an environment allowing an authentic dwelling experience. Heidegger's paradigm of the Greek temple, which occupies space by giving its surrounding space a meaning, expresses a potential that is not lost.

Technoethics tries to find how to build new commons where humans can dwell and flourish. Public administration and urban space experts are ready and willing to implement such places but in my modest experience they are perhaps lacking the most fundamental specifications, which are not technical but technoethical. Once we understand that dwelling is ethically laden and that dwelling is ordinary existential skills, we can certainly weave new threads of agency to reconfigure natural and technological environments. Technoethics' challenge is to help frame the built environment according to a set of values different from the industrial and commercial one that has shaped our world.

References

Allenby, Braden R., and Daniel Sarewitz. 2011. *The techno-human condition.* Cambridge, Mass.: MIT Press.

Basalla, George. 1998. *The evolution of technology.* Cambridge: Cambridge University Press.

Bell, Daniel. 1973. *The coming of post-industrial society: A venture in social forecasting.* New York: Basic Books.

Benkler, Yochai. 2011. *The Penguin and the Leviathan: How cooperation triumphs over self-interest.* New York: Random House.

Bloch, Ernst. 1959. *Das Prinzip Hoffnung, Werkausgabe,* Band 5, 1985. Frankfurt am Main: Suhrkamp. (*The principle of hope.* Cambridge, Mass.: MIT Press. 1986).

Borgmann, Albert. 1984. *Technology and the character of contemporary life: A philosophical inquiry.* Chicago, Ill.: University of Chicago Press.

Borgmann, Albert. 1999. *Holding on to reality: The nature of information at the turn of the millennium.* Chicago, Ill.: University of Chicago Press.

Borgmann, Albert. 2006. *Real American ethics: Taking responsibility for our country.* Chicago, Ill.: University of Chicago Press.

Bostrom, Nick. 2005. A history of transhumanist thought. *Journal of evolution and technology* 14(1). www.nickbostrom.com/papers/history.pdf.

Bostrom, Nick, and Rebecca Roache. 2007. Ethical issues in human enhancement. In *New waves in applied ethics,* ed. Jesper Ryberg. Basingstoke: Palgrave Macmillan. www.nickbostrom.com/ethics/human-enhancement.pdf.

Butler, Samuel. 1872. *Erewhon, and Erewhon revisited,* Introduction by Lewis Mumford, 1927. New York: Random House. www.nzetc.org/tm/scholarly/tei-ButErew.html.

Castells, Manuel. 1996. *The information age: Economy, society, and culture, vol. 1: The rise of the network society.* Oxford: Blackwell.

Castells, Manuel. 1997. *The information age: Economy, society, and culture, vol. 2: The power of identity.* Oxford: Blackwell.

Castells, Manuel. 1998. *The information age: Economy, society, and culture, vol. 3: End of millenium.* Oxford: Blackwell.

Clark, Andy. 2003. *Natural-born cyborgs: Minds, technologies, and the future of human intelligence.* Oxford: Oxford University Press.

Dennett, Daniel C. 1995. *Darwin's dangerous idea: Evolution and the meanings of life.* New York: Simon & Schuster/Penguin Books.

Dreyfus, Hubert L. 2001. *On the Internet.* London; New York: Routledge.

Dublin, Max. 1992. *Futurehype: The tyranny of prophecy.* New York: Plume.

Ellul, Jacques. 1954. *La technique, ou l'enjeu du siècle.* Paris: A. Colin, repr., 1990. Economica. (*The technological society.* Trans. John Wilkinson. New York: Knopf, 1964).

Elster, Jon. 1983. *Explaining technical change: A case study in the philosophy of science.* Cambridge: Cambridge University Press.

Gehlen, Arnold. 1957. *Die Seele im technischen Zeitalter.* Hamburg: Rowohlt. (*Man in the age of technology.* Trans. P. Liscomb. New York: Columbia University Press, 1980).

Grinevald, Jacques. 2008. *La biosphère de l'anthropocène. Climat et pétrole, la double menace. Repères transdisciplinaires [1824–2007].* Genève: Georg Editeur.

Harvey, Peter. 2000. *An introduction to Buddhist ethics: Foundations, values and issues.* Cambridge: Cambridge University Press.

Heidegger, Martin. 1954. *Vorträge unf Aufsätze*. Pfullingen: Neske.

Hershock, Peter D. 2006. *Buddhism in the public sphere: Reorienting global interdependence*. London; New York: Routledge.

Hershock, Peter D. 2012. *Valuing diversity: Buddhist reflection on realizing a more equitable global future*. Albany, N.Y.: State University of New York Press.

Hershock, Peter D. 2014. *Public Zen, personal Zen: A Buddhist introduction*. Lanham, Md.: Rowman & Littlefield.

Hershock, P.D., M. Stepaniants, and R.T. Ames (eds). 2003. *Technology and cultural values: On the edge of the third millennium*. Honolulu: Hawaii University Press.

Hottois, Gilbert. 1984. *Le signe et la technique. La philosophie à l'épreuve de la technique*. Paris: Aubier.

Hursthouse, Rosalind. 1999. *On virtue ethics*. Oxford: Oxford University Press.

Ihde, Don. 1979. *Technics and praxis*. Dordrecht: Reidel.

Ihde, Don. 1990. *Technology and the lifeworld: From Garden to Earth*. Bloomington, Ind.: Indiana University Press.

Illich, Ivan. 1992. *In the mirror of the past: Lectures and addresses, 1978–1990*. London: Marion Boyars Publishers.

Jonas, Hans. 1979. *Das Prinzip Verantwortung*. Frankfurt a.M.: Insel. (Trans. *The imperative of responsibility: In search of an ethics for the technological age*. Chicago, Ill.: University of Chicago Press. 1984).

Jurgenson, Nathan. 2012. When atoms meet bits: Social media, the mobile Web and augmented reality. *Future Internet* 4: 83–91.

Latour, Bruno. 1992. *Aramis ou l'amour des techniques*. Paris: La Découverte. (*Aramis, or the love of technology*. Cambridge, Mass.: Harvard University Press, 1996).

Lovelock, James. 1979. *Gaia: A new look at life on Earth*. Oxford: Oxford University Press.

Midgley, Mary. 1984. *Animals and why they matter*. Athens, Ga.: University of Georgia Press.

Mumford, Lewis. 1967. *The myth of the machine*. New York: Harcourt, Brace, Jovanovich.

Naess, Arne. 1979. Self-realization in mixed communities of human beings, bears, sheep and wolves. *Inquiry* 22: 231–241.

Naess, Arne. 2010. *The ecology of wisdom*, in *Writings*, ed. Alan Drengson and Bill Devall. Berkeley, Cal.: Counterpoint Press.

Nash, Roderick. 1973. *Wilderness and the American mind*. Revised edn., originally published 1967. New Haven, Conn.: Yale University Press.

Negroponte, Nicholas. 1995. *Being digital*. New York: Knopf.

Nisbet, Robert. 1980. *History of the idea of progress*. New York: Basic Books.

Noble, David F. 1997. *The religion of technology: The divinity of man and the spirit of invention*. New York: Knopf.

Nowotny, Helga, Peter Scott, and Michael Gibbons. 2001. *Re-thinking science: Knowledge and the public in an age of uncertainty*. Cambridge: Polity Press.

Nye, David E. 2006. *Technology matters: Questions to live with*. Cambridge, Mass.: MIT Press.

Olsen, Jan Kyrre Berg, Stig Andur Pedersen, and Vincent F. Hendricks (eds). 2009. *A companion to the philosophy of technology*. Hoboken, N.J.: Wiley-Blackwell.

Pirsig, Robert M. 1974. *Zen and the art of motorcycle maintenance*. New York: Bantam.

Puech, Michel. 2008. *Homo Sapiens Technologicus. Philosophie de la technologie contemporaine, philosophie de la sagesse contemporaine.* Paris: Le Pommier.

Roco, Mihail C., and William Sims Bainbridge (eds). 2002. *Converging technologies for improving human performance: Nanotechnology, biotechnology, information technology and cognitive science.* Arlington, Va.: National Science Foundation. www.wtec.org/ConvergingTechnologies/Report/NBIC_report.pdf.

Roszak, Theodore. 1972. *Where the wasteland ends: Politics and transcendence in post-industrial society.* New York: Doubleday.

Savulescu, Julian, and Nick Bostrom (eds). 2008. *Human enhancement.* Oxford: Oxford University Press.

Serres, Michel. 2001. *Hominescence.* Paris: Le Pommier.

Simondon, Gilbert. 1958. *Du mode d'existence des objets techniques.* Paris: Aubier.

Tavani, Herman T. 2004. *Ethics and technology: Ethical issues in an age of information and communication technology.* 3d edn. Hoboken, N.J.: John Wiley & Sons.

Taylor, Paul W. 1986. *Respect for nature: A theory of environmental ethics.* Princeton, N.J.: Princeton University Press.

Thoreau, Henry David. 1985. *A week on the Concord and Merrimack rivers [1849]; Walden or life in the woods [1854]; The Maine woods; Cape Cod.* New York: The Library of America.

Van den Eede, Yoni. 2012. *Amor technologiae: Marshall McLuhan as philosopher of technology – Towards a philosophy of human-media relationships.* Brussels: VUBPress.

Verbeek, Peter-Paul. 2011. *Moralizing technology: Understanding and designing the morality of things.* Chicago, Ill.: University of Chicago Press.

Winner, Langdon. 1977. *Autonomous technology: Technics-out-of-control as a theme in political thought.* Cambridge, Mass.: MIT Press.

Wong, Pak-Hang. 2011. Dao, harmony and personhood: Towards a Confucian ethics of technology. *Philosophy and technology.* www.springerlink.com/content/u667060p46746115.

3 Ordinary Technologies and Ethical Significance

3.1 The Ordinary in Contemporary Life

Philosophy of the Ordinary

A philosophy of the ordinary is possible, and this book tries to show that it gives access to original forms of solace, serenity, and wisdom. The new framework will be demanding for those who still believe that we are on the verge of a complete rational and technological dominance of the world. What I suggest is not only a lateral step; it also entails a fresh look on the entire environment. Instead of the search for the ultimate expertise in the two first forms of power (technology and domination, Table 1.3), I am calling for the search for a humble wisdom, relying on the third form of power – power over oneself as the ordinary agent of one's life (Puech 2013).

The notion of *ordinary* in philosophy originally refers to the two rival schools called "Ideal Language" and "Ordinary Language" in analytical philosophy during the second half of the twentieth century. In the Oxonian version, Ordinary Language philosophy launched a deflationist analysis of the language that is actually used in everyday life. Its goal was to describe the implicit metaphysics of this language. Austin's (1962) dissections of excuses, for instance, are masterpieces of Ordinary Language methods and results. But there seems to be a taboo on substantive ethical considerations in all these quasi-moral inquiries. Or even worse: Ordinary Language philosophy quickly became a sort of aporetic snobbery, proud of its uselessness and indulging in sterile academicism. This aristocratic detachment was abandoned by the posterity of Ordinary Language philosophy when it mixed with American pragmatism. Finally, through the mediation of Dewey's influence on philosophy of technology (Hickman 1990), technoethics resumes the effort to harness the ordinary in a philosophical approach to modern existence.

The attraction of the extraordinary can be explained and must be resisted in our field. Its main force is the disregard for the ordinary, in an ambiguous form. In a volume on "everyday life," Rita Felski remarks:

Everyday life, in other words, is rarely viewed with neutrality. The concept is marked by a rich history of hostility, envy and desire, expressing both nostalgia for the concrete and disdain for a life lacking in critical self-reflection.

(In Highmore 2012: vol. 1, 289)

Among the reasons to explain this ambiguous disregard a prominent one is that everyday life is a realm of women, a "realm" that most of the time consists in what is left after men have taken their part of life's occupations and opportunities, explains Felski.

Gender inequality is not the sole ideology behind ordinary/extraordinary valuation. On the whole, the extraordinary is a fertile soil for ideologies of all kinds, manipulation, and narcissism. The ordinary consists in humble tasks that are deprived of any possible ideological grandeur. When "salvation" is at the heart of one's values system, as it is in Abrahamic religions or in Heidegger, its transcendence needs the contrast with the vile ordinary in order to promote, more or less openly, the "other" world. "Only a God can save us" was Heidegger's conclusion in a famous *Spiegel* interview (Heidegger 1976). The leap into the irrational was not reached in the conclusion: it was prepared by the entire analysis of modernity and technology as "ordinariness," a "fall" into the ordinary, into the sin of the mundane. Some trends in the ethics of technology still follow a scenario disparaging the ordinary in order to introduce the ideological salvation they have in mind. Ordinary ethics has no ideological agenda and thus no prejudicial reason to disparage the ordinary. There are good reasons to celebrate it instead.

Quine's book *Word and object* (Quine 1960) gives an extremely stimulating account of the formation of the human mind, in ordinary experience, from sensory stimuli up to language and logic. Quine's perspective is "a logical point of view" but also a naturalistic point of view. The ontology of ordinary objects has taken the path of logic and formalization (Hickman 1988; Baker 2008; Korman 2011); my intention is to explore the path of naturalization. Quine's model highlights the role of ordinary acquaintances in the "wiring" (half a century later we can be more affirmative in neurology) of our cognitive skills and meta-skills. The essential function of language and logic is to support the acquisition and practice of other skills. Among the skills acquired by ordinary acquaintance and practice we need to make room for practical, physical, artifact-related, and (now) digital skills. Technoethics intends to elucidate the functional and psychological merging with artifacts and to analyze the emotional and ethical bonding with the technosphere and infosphere.

Quine's first paragraph was entitled "Beginning with ordinary things" (Quine 1960) and it is actually as fresh a beginning in ethics as it was in logic at that time. From ordinary things the human mind weaves a web made of idiosyncratic pathways that build the inter-individual common language and a common world. As Quine brilliantly put it:

Different persons growing up in the same language are like different bushes trimmed and trained to take the shape of identical elephants. The anatomical details of twigs and branches will fulfill the elephantine form differently from bush to bush, but the overall outward results are alike.

(Quine 1960: 8)

A common language and a common ontology are based on the natural growing (of neural connections) in the ordinary experience of the world. There is no reason to believe that emotions and values are not following the same process.

For Quine, an *ontological commitment* is the acceptance of the existence of a thing of any kind. A central Quinean tenet is that assumed existing entities (material or not) are by definition the possible values of a variable in a logical sentence (Quine 1953: 1–19), that is to say something mentioned in a sentence. For instance, if I say "I can lend my car to you tomorrow," I ontologically assume that I do have a car and you can ontologically assume the same fact. It seems quite natural to extend this definition to assumed *values*. The same example works: if I say "I can lend my car to you tomorrow" there is a *moral* commitment on my side to have a car and a *moral* commitment to lend it to you if you want. I morally assume a certain number of values pertaining to consistence (about owning and about lending) and obligation (about lending). This is a case of "performative" talk in a promise, according to Austin's theory of "doing things with words" (Austin 1962).

More generally, endorsed values are values "mentioned" by actions in a wide variety of manners. The extended modes of *mentioning values* can be linguistic, either in justification of the action or in the deliberative reflection about it (motivation). Values can also appear in the commentary on the action offered by external observers, especially a neutral and benevolent third party observer. Values can also be metaphorically "mentioned" by actions, evoked by them, even in the absence of any internal or external discourse containing the appropriate linguistic descriptors. An act of courage, particularly of ordinary courage (resisting ordinary domination), is an instance of courage (the virtue and the value) even without any linguistic accompaniment.

There is an interesting consequence of this Quinean line of thought: values can be *inscrutable* in behaviors, exactly as references are inscrutable in Quine's philosophy of logic and for the same reason (broadly construed) of originating in the constructive skills acquired in the ordinary experience of the world. What takes place in ordinary experience has a status similar to the "transcendental" structures in Kant's philosophy: constructive structures, meta-structures or meta-skills assemble the framework in which everything is framed (things and values), a framework that is not in itself empirically observable.

Jean Piaget's research on the stages of children's development is also relevant here. Piaget's research embraced the cognitive as well as moral development, including the progressive and interactive acquisition of logical and moral skills (Piaget 1932: 322). Lawrence Kohlberg's works (Section 1.2) and Francesco Varela's neurological approach (Section 4.2) reinforce this theory. Ancient Chinese wisdom has noted the fact: "The little child learns to speak, though it has no learned teachers – because it lives with those who know how to speak" (Zhuang Zi 1968: 299).

It is an Ordinary World

We are living in a world of ordinary entities and micro-behaviors. Philosophy can take a fresh start from elementary consumption actions and also elementary production actions, within this level of ordinary life. The primacy of the end-user, one of the most important features of the modern world, is rooted in the ordinary existence of this end-user. Historians have emphasized the importance of "material civilization." My focus is on material civilization, the mundane civilization, made of artifacts and devices, technological systems, more and more interconnected – the Internet being the extreme achievement of this material (now largely digital) civilization. The intricate networks of technological systems for production, transformation, transportation, energy, shopping, security, information processing and retrieving, and so on, all this fits in with the definition of a "material civilization" in the spirit of Fernand Braudel's original notion.

I suggest that one's mixed feelings for one's mobile phone can be even more complex and significant than one's religious feelings. Rather than "eliminating" ordinary objects, a common practice in Ideal Language philosophy, I am inspired by Ordinary Language philosophy in trying to understand how we manipulate ordinary objects, how we evaluate and value them, feel about them, symbolize through them.

Michel de Certeau and his team examined life in modern French cities, in its ordinary dimensions like habitat, cooking, traditional shopping and socializing in the streets, shops, and markets (De Certeau 1980; De Certeau et al. 1980). This style of analysis has been very influential in sociology all over the world. Once it is rid of the jargon used by sociology in the 1970s it is still inspirational to address the current ordinary lifeworld. Its emphasis on a popular "resistive" culture, in the style of Richard Hoggart's research in England (Hoggart 1957), remains a reference for characterizing the resistive reappropriation trends in our high-tech culture.

A differential analysis of the *ends and means structures* of ordinary technology is appropriate here. What about our actual use of the smartphone as a "means"? My last email checking while listening to music yesterday were very superficially (hypocritically?) "means" to care for the people that I love and the varied business that I run; they were more sincerely little moments of self-comforting, by checking that the technology still works smoothly and remains obedient at my fingertips. Our incessant micro-uses of domestic appliances follow the same path. Preparing an espresso coffee at home by just pushing a button is in itself a comforting praxis. Sometimes the technological experience brings more real comfort than the hot cup of coffee itself (which in some cases may not even be drunk, it is forgotten somewhere). Smoking a cigarette is a dangerous version of this experience; getting money from an ATM in the street is another: means and ends are not so clear-cut and the comfort of the technological technosphere in itself is deeply embedded in these experiences. Cooking with modern tools, even a simple ceramic knife, is a complex pattern of means and ends, still implying a multitude of "traditional" factors but now suffused with technology.

There is even more: the fact that these technologies are ordinary and transparent is an essential part of their existential inscription. The familiarity of smartphone checking and push-button espresso is a significant comforting experience of the technosphere, our lifeworld. My emphasis is not on the shallow impression of comfort given by ordinary technology, even if I do not despise it at all. It is on the existential depth of these "world checking" actions: an uninterrupted reassurance flow runs between us and the technosphere in our ordinary life.

Ordinary Technology Familiarity

Ordinary familiarity with technology constitutes the background of modern life (Lie and Sorensen 1996). We are surrounded by available commodities and resources: emergency phone calls at any time and place are possible with the mobile phone, drinks or snacks are available everywhere in industrialized cities, almost any information is "one click away" on the Web. Assessing "Internet in everyday life," Maria Bakardjieva explores interpretations of the fact that the Internet is ordinary and not extraordinary (Consalvo and Ess 2011: Chapter 4). She refers this experience to Highmore's (2002) analysis of "everydayness," the value and specific quality of the everyday.

Trying to remember the name of an actor can take minutes, finding it online will take seconds: there is no mystery if our exosomatic memory is a familiar existential experience. Consequently, the critique of commodification may not be the last word in the interpretation of technological familiarity. Borgmann (1984) tends to suggest that commodification is the way things and devices are made familiar to us. A less subtle social critique would point directly at the marketing conspiracy behind that commodification process. No doubt the questionable aspect of commodification is its inauthenticity and Borgmann's program to restore authenticity is also my program. But the only way to correct this perversion may not be the restoration of traditional family life, as advised by Borgmann. A broader and innovative restoration of authentic familiarity, intimacy, and reassurance bonds with persons, living entities, and mundane objects must be on our agenda.

Historically, the familiarity with electronic artifacts can be traced back to the 1970s, when Asian mass-market electronics devices flowed over the

West and then the whole world. Sony's Walkman paved the way for today's mp3 devices and more essentially for "wearable" devices, a promising ecological niche. The interesting moment is the moment when wearing earphones in public space (street, transportation, park, restaurant) is no longer perceived as an anomaly but as a normal, ordinary behavior. Another interesting moment is when wearing earphones at home (but not during the family meal) or at school (but not during classes) is perceived as normal and becomes an ordinary feature of life. A complex web of familiarity sets in, very quickly. It entails psychological changes for the self (for instance the presence of a musical background that separates the self from the immediate environment) and social consequences: a conversation with earphones still on, or the effort to pull them off when entering into a casual conversation are new features of modern social interaction.

Technoethics bears on the micro-behaviors through which users appropriate technologies. The constant flow of innovation and design supplied by the industry provides the variants that are submitted to a Darwinian selection in real life through a complex pattern of collective and individual microactions. Innumerable tools and devices in the technosphere are *appropriate* to their functions (material and symbolical) because users have *appropriated* them through real use. The short but intense history of personal computers gives evidence of rapid adaptation and feedback on both sides, object and subject. Coevolution is a story of co-adaptations driven by the most ordinary use. None of the basic microcomputer users have become computer scientists. They do not even need to know the MS-DOS syntax that was necessary in the 1980s for PC users because digital interfaces have been remodeled under the pressure of millions of ordinary users through their capacity and incapacity to learn, understand, and remember; through their micro-acceptances and micro-refusals. Some actual uses are aberrations, for instance to systematically access websites through Google's search bar or through Facebook (rather than the URL bar of the browser or bookmarks). I do not say that these aberrations and their harmful consequences are positive. On the contrary, they are wrong appropriations and uses. Technoethics here clearly dares to be normative.

The constant effort of designers to improve the usability of devices does not imply that the designer is the final prescriber of usability improvements. For successful companies, real uses and actual appropriations are the empirical sources of product improvement, whereas neglecting this monitoring of uses explains fairly well the misfortunes of technology deployment in bureaucratic institutions. The primacy of use and appropriation is not only a factual element in the technosphere; it qualifies as a value in technoethics. When innovation is imposed on the user, when appropriation is forced by a dominating power, the moral value of ordinary empowerment is impinged. Moral micro-abuses of this kind happen all the time in the workplace where IT bureaucracy rules and sometimes also in public space where national or international IT governance or corporations try to prevent ordinary users from doing this or to oblige them to do that.

The primacy of use and appropriation obtains in the lifeworld at large, beyond high-tech products. Running shoes, for instance, transform the experience of one's body and physical capability and invites exploring the environment and one's bodily resources. Mindfulness when taking a hot shower is an experience of ordinary appropriation that goes beyond functional utility and reaches existential significance. The existential experience of driving a car must be understood as one of these deeply satisfying appropriations if we want to address the addictive attachment to the car.

The most important empirical evidence to be retrieved from the field today is certainly about screens. Three screens configure the contemporary lifeworld: television, personal computer (including tablet and game console), and mobile phone (especially smartphone, Internet enabled). The history of recent modernity gravitates around these three screens, their use and appropriation, and of course the huge social, political, and economic structures related to these screens. The most significant coevolution here leads to an ever-closer proximity to the body and the self. Motion pictures were at first a distal event, happening at a distance, on a wall-screen in public and institutional occasions. Then the TV brought the show at home, nearer and nearer when TV screens penetrated every room in the house. The computer screen followed the same path and the present "laptops" and tablets are proximal artifacts in the immediate vicinity of the body, in constant touch, under the eyes and fingers of the self.

But nothing compares to the *wearable screen* of today: the smartphone. The use and appropriation of the smartphone are fundamental for ordinary technoethics. They exhibit all the dimensions of the existential significance lately bestowed on ordinary digital technologies. The significant phenomenon with the smartphone is not the extraordinary technical accomplishment of connecting anyone to anyone or anything in real time through an autonomous device the size of a packet of cigarettes – which nonetheless remains a technological prodigy. It is the penetration of this artifact in the lifeworld. The speed and depth of this penetration is amazing and unprecedented. Even in North Korea, one in ten inhabitants owns a (totally illegal) smartphone, according to a 2014 study (http://uskoreainstitute. org/wp-content/uploads/2014/08/Kim-Yonho-Cell-Phones-in-North-Korea-English.pdf). "Mobile telephony is the fastest diffusing communication technology in human history," remarks Manuel Castells in the conclusion of an edited book surveying the full range of mobile diffusion (Katz 2008: 447). These studies confirm the sudden social importance of mobile phones in less industrialized countries and the reconfiguration of social organization in every kind of contemporary society, not as a planned social engineering transformation, but as a consequence of ordinary adoption. In the avalanche of research about the social disruptions due to mobile phones, the positive aspects are linked to ordinary interpersonal connection and micro-coordination (Ling 2004). A new density of *personal permanent synchronization* is now essential in everyday life.

When a plumber or another craftsman comes to your home today he/she will be equipped with a smartphone which is used as a torch, a camera to take a picture of the installation, an agenda for planning the intervention, a stock management server for the needed replacement parts, a calculator for the cost estimate and invoice, which can be sent by email on the spot. In a lot of cases this smartphone is not a professional tool, paid for by the employer, but a personal device, typically a Christmas present. It is not a traditional professional tool and yet nothing can be more professional. It belongs to the ordinary technosphere – where professional life draws its main resources, even for plumbers.

Immersed into the Digital, Surrounded by the Smart

The social dimension of the ordinary is still investigated most of the time with twentieth-century methods, aiming at social critique through the disclosure of hidden determinations and vile hidden agendas. Technoethics stands on a different footing as it tries to consider ordinary social life as constitutive (determining) and not only as constituted (determined). The principal difference is that our method returns to the self a real social agency in its everyday informal connections to other selves and social structures. The turning point is the digital empowerment of the self in the twenty-first century.

The ordinary of today is not exactly a trivial ordinary since it is immersed in the digital. Pervasive digitalization has transformed human existence in the twenty-first century to the point where the ordinary is no longer the lower stage of social structures but rather the entry interface for social systems. Every detail in ordinary life (How can I fix my washing machine? Should I pay attention to this strange new spot on my skin? Is it going to rain this afternoon?) is potentially propelled to satellites and huge computers and it bounces back on my screen in a microsecond. The density and responsiveness of the infosphere is a radically new property of our living environment and it configures at low level our dwelling in the technosphere. Through high-tech and small-tech devices (Hawk et al. 2008), in the constant presence of the infosphere and in permanent connection, our ordinary life retains the typical humanizing potential of any ordinary technosphere in any civilization, but it also transcends the limitations of any previous life environment and it explores radically new potentials. Once again, the philosophical assessment of this epoch depends on the "normalization" of this experience: it happens in the ordinary.

The notion of "smart" is a key feature of this new phase in ordinary technologies. Poor design was frequent in technology, including the first generations of digital technologies destined to the public. From some audio tape recorders and other domestic appliances it could once be said: "You would need an engineering degree to figure out this" (Norman 1988: 1).

"Smart" devices have changed and even reversed this "psychopathology of everyday things" and the "frustrations of everyday life" (Norman 1988). Smart objects diffuse a sort of user euphoria that gives a new tonality to the device paradigm. The *smart-device paradigm* describes user experience when accessibility is transparent. The device is not even seen as the access to something but as the immediate presence of it. User-friendliness is not limited to understanding my needs and intentions but it anticipates my needs and intentions. When the "app" is smart enough to be transparent, a remarkable fusion with the self's stream of consciousness takes place. I "am" in the virtual bookstore and I "grab" the desired book. The infosphere zone that I am projecting myself toward (a perfect phenomenological case of digital "intentionality") is reached by a digital avatar of my self. This avatar is a part of myself: it learns and buys for me, it talks for me, it is me as an agent. My point here is to stress the fact that this merging into the infosphere is an extraordinary experience occurring every day.

Smart ordinary devices have accomplished the "singularity" that has long been expected from Artificial Intelligence. But it happened at another unexpected scale, the scale of the ordinary (everyday utilities) rather than the extraordinary (institutional power and mega-machines). "Smartifacts" (Ma et al. 2005) must be celebrated for this technoethical decisive shift: from a (miraculous) governing, dominating, and controlling global intelligence, which was expected and feared from AI, toward the now existing ordinary service-driven local smart devices – small and affordable, irresistibly user-friendly. The "wow! effect" caused by the first use of an application like Shazam (www.shazam.com), which instantly identifies any piece of music, is just one example of the accomplishments constituting the smart.

Three characteristics of smart devices are remarkable. All of them are prominent in the smartphone, which is clearly the smartest device of today.

1 *Convergence* makes the smartphone a real Swiss Army knife since it is the best telephone, pager, short text sending device (including short emails), emergency call lifeline and global safety device, camera (for fun or for recording police violence), radio receiver and mobile music player, GPS and map of everywhere, public transportation timetable and real-time tracker, notebook, appointments calendar and reminder, alarm clock and wristwatch, dictionary and language assistance service, portable encyclopedia, portable video-game pad, voice recorder, short text reading device (news feeds), mirror (front-camera) for make-up or detecting salad remnants between teeth, etc. It is second (to a good laptop computer only) for: long email or long text writing/reading, Web browsing (including online shopping), theater-like film experience, etc.

2 *Wearability* and pocket-wearability (like glasses, wristwatches, keys, USB sticks) make smart devices the key to mobility, one of the

ordinary dimensions of the modern self and the modern technosphere. The technosphere became ordinary as it was transformed from furniture to gear: when you move a desktop computer you move the furniture; when you move with a laptop computer you move with a small piece of furniture in a bag; when you move with a smartphone you just move, with clothes on and having your life-gear with you.

3   *Technological transparency* is essential to the ordinary experience of smartness: the phone shifts from 3G to wifi, from app to Web, from SMS to email, as transparently as it shifts from one relay emitter to another, from one corporate- or public-owned network to another, from one version of its OS to an updated one. This transparency requires a critical awareness, but this is another issue.

The most important and specific use of mobile phones, according to empirical and conceptual studies, is "texting," that is to say SMS (Short Message Service). It perfectly exemplifies the new importance of the ordinary in the technosphere. Texting is a user-invented service taking advantage of a narrow purely technical (originally) communication facility which was implemented in the protocols. Its success was not anticipated by designers. It is a low-tech application inside a high-tech environment. It is virtually free and unlimited. It works on the principles of opt-in real-time – I can have real-time communication but only if I opt in, that is if I continuously check, read, and answer my SMS messages.

3.2 The Ethical Importance of the Ordinary

The Ordinary as Ethical Commitment

The ethical significance of ordinary technology follows from the existential importance of the ordinary in technology. Technoethics is rooted in our specific ontological commitment to ordinary technology. In each and every case this existential commitment is laden with an ethical engagement. *Importance* is the key ethical notion in the approaches that can be fruitful for technoethics, especially the ranking of concerns and things we care for (Frankfurt 1988). Everyday importance ranking and ordinary caring for do not refer to transcendent truths; they belong to ordinary behavior – now mediated by technology.

Ordinary instances of value are elusive but crucial. What I decide to eat for my next meal; the person I decide to call on the phone, or not; the authorization I give, or not, for an application to track me; the second I take, or not, to read and understand the label before buying or not buying: all these micro-behaviors are ethically laden and they are in fact ethically decisive. They are the actual ethical decisions in our world. In these instances a value can be "mentioned" explicitly but it is rarely so. However, buying an extremely cheap item "made in PRC" in fact *mentions* a value: the acceptance of underpaid and possibly forced labor in

China or the laziness involved in not understanding what "PRC" means. Forwarding a compromising email for fun or for revenge in fact *mentions* a value: the low respect for persons and privacy when technology makes it easy to be nasty. The values mentioned by actual behaviors in these instances are negative but positive values can be mentioned by ordinary behavior as well. Ethical commitment in the ordinary often consists of mentioning values by one's action, which is the perfectly normal and sound practice of values and virtues – in Confucian or Buddhist philosophy more than in the Western logocentric conception of moral life. In the West, the exclusive value of universal truth and linguistic representation (be it religious or scientific) is linked both to the prevalence of technology (domination over things) and to the modern institutional state (domination over people), the two first kinds of power (Table 1.3). The focus on ordinary microactions and their decisive ethical significance radically changes the outlook and opens the way for the third form of power (wisdom). Technology fits perfectly within this new perspective because it belongs to the realm of action and not the realm of discourse. The shift from the ethics of (true and general) discourse to the ethics of ordinary microactions is an essential move to establish technoethics.

The ordinary is a rich source of wisdom in Ancient Chinese philosophy and in Buddhism. Ames and Hall (2001) explore the ethical lesson of the Zhongyong's central doctrine, "focusing the familiar" – the Chinese term *zhongyong* is astonishingly close to Borgmann's "focal," it conveys the ideas of central, familial and familiar (Ames and Hall 2001: 43). Hershock insists on this philosophical pillar of Buddhism: "ordinary mind is Buddha" (Hershock 2014: 69). And the Japanese philosopher Watsuji puts it so: "[…] there is no mine so rich as that which is called the 'everyday experience' of human beings" (Watsuji 1996: 39). Interpretations of modernity as a "fall" in mundane everydayness have missed something, which can be reclaimed.

Disregard for the Ordinary: Modernity as Dereliction

The importance of ordinary life in the philosophical assessment of contemporary technology has been formally recognized in recent philosophy of technology, essentially under the influence of Borgmann (1984). This is true only for the openly ethical approaches, particularly those including the question of the good life (Higgs et al. 2000; Brey et al. 2012). Nevertheless, there remains an ambivalence in the treatment of the ordinary in philosophy. The "promise of technology" was not about ordinary life only, it included space travel and object teleportation, but what has been delivered on that promise is about ordinary life: "home cinema" systems to watch SF fictions or websites to buy things and have them at home at the end of the day. The ambivalence is not about downsizing expectations. Space travel as it was fantasized, in the form of weekends on the moon or space wars, is in reality less important than the cultural potential of home cinema. The dematerialization of cultural contents (books, music, cinema) on the Web already equates a teletransportation of objects, for a marginal cost tending to zero. The disappointment with the "ordinary" delivery on the "extraordinary" promise of technology is largely a perspective illusion. We are driven to it by the undervaluation of the ordinary, which is but one of the aspects of the largely technophobic evaluation of modernity in philosophy. This evaluation has been challenged, however. Against the common interpretation of modernity as "disenchantment," Jane Bennett for one is drawn to "the enchantment of modern life." She gives examples of ordinary moments and ordinary artifacts, such as the computer (Bennett 2001: 171), that induce a sense of elation and a lasting attachment. Although she affirms that there is an ethical potential in this ordinary wonder and in ordinary attachment, her post-modern method is prone to literary enrichment of the theme more than listing constructive philosophical assets.

In Heidegger's seminal analysis of the *Dasein*, the ordinary is accurately described and understood as a specific existential level, a global structure of our relationship to the world. For him, an obfuscating view prevails in the ordinary: objects are seen as instruments, time and space are seen as the environment of instrumental projects, and our own self and its potentials are seen as hollow impersonal entities (Heidegger 1927: Sections 9–38). Everydayness is a primordial concept but here it means dereliction, the loss of the *Dasein's* transcendent status, which is an ethical status – even if "transcendent" and "ethical" are not correct in Heideggerian terminology.

The influence of this masterful analysis on the subsequent existential approaches to technology has been and still is devastating. My position implies a reconnection with its inspiration and a redirection toward a *constructive* interpretation of everydayness and the ordinary. Heidegger's verdict on the inauthenticity of mundane preoccupations, which leave no time for real existence, remains a solid foundation for an ethics of modernity. Accounting for the ethical significance of the ordinary implies a strong concern for the inauthentic preoccupations that deprive us of real life. The question, however, is double. Is this inauthenticity a universal feature of modern technology? Is this inauthenticity an effect of these technologies? The first question is about technophobic essentialism and the second one is about technological determinism. Heideggerians tend to adopt these two doctrines, even if it is in sophisticated versions. I suggest relinquishing them, in any version.

Let us take the utmost ordinary and utmost despised technological device: TV. "Seeing how television both enthralls and disappoints people, we may have a clear view of what flaws technology at the center," says Borgmann (1984: 125). It may not be so one-sided. When television programs were entirely made by the state and by corporations, with a reduced and illusory choice between channels and programs, then control and stupidification (we need this neologism) power were a central reality in personal and social life. But in less than a generation what was televised became less and less one-dimensional. For instance, TV reports about the Vietnam war were a main cause of the public opinion change of mind about waging this war. Active and even resistive attitudes are always possible, even from the couch in front of the TV set. The approach to ordinary dereliction should focus on the derelict self and not on the technological device involved. An ethical meta-analysis would tackle the issue differently, focusing not on the stupidity of people, nor on the stupidification power of the media, but on a meta-level: on the meta-attitude of non-resistance, acceptance, the lack of self-consistence, self-reliance, self-care. TV is an ordinary and ethically very significant element of contemporary life, I agree. Its consequences on human values are mainly negative and they broadcast inauthentic experiences of the world, I agree. My point is to concentrate on what lies between the screen and the couch, as people say in computer technology: "the problem is between the screen (or keyboard) and the chair." It means that the human is the problem.

To continue the rehabilitation of ordinary technology let us give a closer look at Borgmann's critique of commodities. In Heidegger there was no real positive celebration of the authenticity that contrasts with modern dereliction (except allusions to Romantic poetry and, alas, veiled allusions to extreme right politics). In Borgmann modern dereliction is clearly contrasted with *focal* things and activities that carry positive values, the values of human flourishing. "A focal practice is one that can center and illuminate our lives" (Borgmann 1984: 4).

In Borgmann's device paradigm, meaning is lost because context is lost:

> Devices, that was the claim, dissolve the coherent and engaging character of the pretechnological world of things. In a device, the relatedness of the world is replaced by a machinery, but the machinery is concealed, and the commodities, which are made available by a device, are enjoyed without the encumbrance of or the engagement with a context.
>
> (Borgmann 1984: 47)

Advanced devices work in a sort of functional isolation, which in the end is the solitary and meaningless act of consumption. Let us take examples from the family kitchen, a place where Borgmann situates a lot of the focal and where, as a Frenchman who cooks, I may claim basic competence. If coherence and engagement only characterized the "pretechnological world of things" there would be no place for it in the kitchen, because cooking is entirely technique and technology, by definition. Raw meat and roots may be food but not cooking. As soon as you have a fire, pots and bowls, and a variety of ingredients, you can process according to recipes (culinary programming). After some millennia we currently use an electric convection oven (with electronic temperature sensor and programming capacity) and a gas stove (electronically driven today), and our cooking process includes some frozen raw material (that may consist in mushrooms personally picked up last fall in the wild). Contexts, meanings, and engagement: are they really lost? There is a plausible loss of meaning and value, but from the point of view of a total stranger. Say a stranger comes from the megalopolis and visits you in your Montana or Pyrenees' retreat. He/she pops up in the kitchen and is horrified to see that you are dialing a program on the oven keyboard and that the mushrooms come from your deep-freezer shelves. While absorbed in the focal activity of cooking, you do not feel the alleged dereliction of modernity the same way.

In the 1950s and 1960s, it was customary and perfectly normal to go for a car ride just for the pleasure and entertainment of it, going nowhere, just enjoying the pleasure of possessing and driving a car. This behavior appears today at the same time (environmentally) criminal and (socially) stupid. But "from where," as a Nietzschean would ask, do we allow ourselves to be so judgmental? Certainly not from inside the psychological and social context of this episode, which was clearly experienced as a particularly *focal* activity. The car ride with family and friends was a celebration of peacetime and of consumption (meaning something in Europe after war and after post-war restrictions), a living symbol of social upward mobility (still functioning at that time), and an exploration of the neighborhood. This kind of activity, in some times and places, for certain people, involved meanings, means and ends, that the out-of-context observer would hardly grasp from his armchair half a century later.

The New Ethical Importance of the Ordinary

In the accounts of language and logic skills learning in ordinary experience and communication, from Piaget to Quine, there is an ontological rehabilitation of the *vernacular*, an important form of the ordinary. We must understand the vernacular in the militant meaning of this notion, due to Ivan Illich:

> Vernacular is a Latin term that we use in English only for the language that we have acquired without paid teachers. In Rome, it was used from 500 B.C. to 600 A.D. to designate any value that was homebred, homemade, derived from the commons, and that a person could protect and defend though he neither bought nor sold it on the market. I suggest that we restore this simple term, vernacular, to oppose to commodities and their shadow.

> (Illich 1980)

The ordinary as vernacular is as basic and substantial as language and logic but it extends to the entire range of domestic skills, in which I will naturally include not only domestic appliances but Web usage and smartphone skills. For an under-gifted player, an electronic guitar tuner is part of the focal activity of music playing in family and friendly circles, a

Borgmannian case obviously. With the perfectly tuned guitar the under-gifted is more at ease than before with an approximately tuned instrument. This comforting device, the electronic guitar tuner, involved in a focal activity, is a tool for conviviality. It belongs to the vernacular in its extended sense, even if it is not "home-made" – the guitar is not home-made either, but the music is.

Borgmann recently put forward the idea of "real ethics" (Borgmann 2006). In this perspective, reconsideration is possible for "the weakness and the competence of ordinary people" (Borgmann 2006: 141) or "the economy of the household" (Borgmann 2006: 161), a domain that was ethically primordial in Emerson as well. But Borgmann's treatment remains exclusively devoted to the denunciation of commodities. At a remarkable moment, he touches upon the acceptance of the ordinary that I try to defend when he concedes: "There are always occasions where a Big Mac, an exercycle, or a television program are unobjectionable and truly helpful answers to human needs" (Borgmann 1984: 208). But the existential strength of these occasions is due to their "ordinarity," a property that philosophical literature rarely evokes. Yet, the real level of the ordinary can be illustrated with Borgmann's two favorite examples: cooking and running. Spending hours in the kitchen to prepare a meal from fresh and raw products is no longer an ordinary and everyday activity. For ordinary people in industrialized countries ordinary cooking means fifteen minutes (or less) preparation with already processed ingredients. In that time and with that sort of ingredients the contemporary human being can ruin or consolidate his/her health; develop or impoverish the culinary taste, habits, and culture of the household; make it necessary or not to earn more money for food; enrich or strain some corporations; help or prevent sustainable farmers to make a living; devastate or protect natural landscape, and so on. The moral register here is different from the values in the preparation of a festive extraordinary meal. The same difference exists between running a Paris or New York marathon: everyone you know is aware of the fact and admires you, photos and movies will show you on Facebook and, with a little luck in newspapers or even TV, your brand new and high-tech equipment does more than keeping up with the Joneses. Ordinary running is different: it happens in your neighborhood, with minimal equipment and maximal discretion, especially if you take it as a meditation exercise. More focal than anything else in its simplicity, ordinary running is authentic and this makes the difference. The snobbish runner is looking for the (illusory) extraordinary.

My point is not only to say that *there is* ethical value in the ordinary, but that the ordinary *is in itself* the key feature of a new kind of ethical value. A new ethical framework is emerging from this consideration. In this framework, the ordinary use of a smartphone bears ethical values that the patronizing anti-smartphone militant totally misses – even in the rare case where the anti-smartphone bragger is actually not a smartphone owner or user, provided some casuistic exception for himself/herself. But conversely the fashion-victim whose arrogance is focused on having the last model of iPhone totally misses the ethical robustness of someone using an "old" mobile phone – provided the voluntary simplicity of the latter remains a private matter (neither bragging nor patronizing). Thus, there are ethical differences in ordinary and personal technology, but their ethical significance is not in the device, it comes from the person.

Use, Affordance, Commodification

The primacy of use and appropriation has consequences in the ontology and functional analysis of the technosphere. We use screwdrivers to open paint cans (what else can be used?) and we use for pure fun or trivialities a lot of high-tech devices that were designed for serious work. The computer to watch pure entertainment movies or to run a video game is the same as the computer for professional work. A first step toward a technoethics of the ordinary is to understand that these non-standard "use plans" belong to the philosophical category of *actions* and not of *functions* (Houkes and Vermaas 2004). But inserting action into a new "metaphysics of artifacts" has philosophical consequences that invite a second step – the ethical one after the ontological one. Ordinary alternative use plans are ethically significant and not only functionally important. Taking them into account impacts the design of artifacts and reorients the assessment of technology. Quite a lot of smartphone uses and Web practices are alternative use plans and a specific set of values obtains within them.

The emerging notion of *affordance* helps apprehending this phenomenon. As is typically the case with new concepts, Wikipedia's page on "affordance" is the best entry gate to the notion (http://en.wikipedia.org/wiki/Affordance). The affordances of an object are all the actions that it allows to its user, particularly the actions that it suggests by its appearance or some of its features, but also all the actions that are discoverable in the course of using the object and in the broader sense of the term all the actions that are (theoretically) possible with it (Gibson 1986). With a simple object like a screwdriver the list goes from regular screwing to opening paint cans, then the use as a hammer or even as a weapon, and the myriad of uses a tinkerer (not a pejorative term) can think of, in the heat of the action.

With smart artifacts, affordance reaches new heights. The more or less unlimited affordances offered by the Web and digital artifacts in general characterize the present technosphere. The design of smart artifacts tries hard to display apparently unlimited affordances. Affordances are what we want and what we want to buy: artifacts that enable us to find our way to the hotel everywhere in the world (even if we do not travel much and if a city map is easy to find); artifacts that enable us to buy and sell on the stock market instantly (even if we lack the money for such a pastime); artifacts that enable us to manage millions of calculations (even if the management of our bank account requires one subtraction a day and one addition a month). The picture may have something ridiculous in it but not from the point of view of the ordinary technosphere. The dozens of items in our wardrobe provide comfort and reassurance even if some of them are never worn in a year. A philosopher unable to understand what a pair of shoes that will never be worn can contribute to a human person is off to a bad start in technoethics. This kind of symbolic, psychological, emotional, and globally existential experience with artifacts is best understood as *affordance* than as use. Possible use exceeds by far any actual use; affordance exceeds possible use.

Decommodifying ordinary devices is possible, to go further in the rehabilitation of the vernacular. The turning point is the interpretation of comfort. The main affordance of the technosphere is comfort, a notion that is not necessarily restrained to a pejorative bourgeois lifestyle. The grandeur of comfort can be regained from German philosopher Arnold Gehlen's anthropology of modern technology. He calls *Belastung* a "load," the ordinary chores of existence, and *Entlastung*, facilitation as "reduction of the load," the relief due to technology (Gehlen 1957). Human existence is constitutionally *loaded* and technology *unloads* it. The philosophically suggestive double sense of "comfort" in English (Oxford English Dictionary online) helps restore the notion in its dignity: "1 – a state of physical ease and freedom from pain or constraint; 2 – consolation for grief or anxiety." Both are human values of importance and they do not essentially refer to bourgeois home furnishing. Comfort is an acceptable keynote notion in technoethics. Its comprehensive apprehension gives way to a particular moral acceptability for affluence. Comfort is an opportunity; it must be conceived as a means and not an end. It does not achieve any value but it puts the person in the right position to engage with ethical projects.

In a resistive decommodification of devices, the ethical significance of the ordinary is exactly the opposite of the currently assumed alienation by commodities. It happens when, for instance, an Internet enabled device that was bought under the influence of marketing push is used to access information sites that are not controlled by trade corporations. This is even more the case if the device is used in alternative circuits for transportation, exchange of skills, services and second-hand products, and why not some political alternative action? Social resistance movements everywhere now typically *hack* technologies that were launched as innocent commodities.

When vernacular skills grow into an alternative lifestyle, they bring about the ethical foundation for new forms of social action. This new form of post-political activism does not consist in ideologically rejecting commodities but in reappropriating artifacts and practices. Thus, the ethical significance of the ordinary goes beyond the acceptance of comfort and facilitation; it dovetails into a radical assessment of the infrastructures of comfort and facilitation. The critical assessment comes from the inside and it takes advantage of the new potentials in the technosphere. With a computer, one can access the Web and set up a devastating critique of computer corporations or of the hidden agendas that are trying to control politics and the economy.

3.3 Awareness and Ordinary Virtue

Awareness as a Virtue

To construct a robust definition of ordinary wisdom the virtue of awareness must be put forward as the adequate mode of consciousness, of being-in-the-world, for the contemporary self. Most of the previous arguments cleared the way in order to access the ordinary in its existential significance and, as such, they were advancements in awareness. The right metaphor for the awareness of the ordinary is not the image of going "deeper" into modern life or environment. On the contrary, in a typically sapiential style, what matters is the difficulty to see what is near, so near that it is transparent. The social researchers who spend their working day stuck on their computer while writing against computers (I know dozens of them) are simply not able to *see* themselves. They are engaging "deeper" issues, they say. We Westerners are not able to see ourselves in the affluent material and cultural environment we live in. We have "deeper" concerns. Awareness as lucidity is the capacity to stay at the surface, without indulging immediately in explanations and justifications, commentary and defamation of the superstructures, of the invisible and visible hands, of all the machinery "behind the scene." Both in phenomenology and in Eastern thought the *candid look* is a precious capacity to reinitialize one's perspective. In this sense, awareness of the ordinary is a candid and even naive practice – and it is also a meditation practice in the process of self-construction.

As a constant effort and practice, awareness is not an independent virtue and it is certainly neither an end nor a value in itself. Consciousness is valued in Western cognitivist philosophy on the basis of its assumption that the access to anything (not only the truth but also the good, values, justice, and wisdom) implies cognition and representational knowledge (Rorty 1979). Awareness is taken in this book in a pragmatic and pragmatist way, not as representational data to be processed but as an existential experience and practice. It prepares wisdom and remains one of its permanent exercises.

The dangers of the contemporary technosphere are said to come from the irresistibility of the "one best way" always offered by technology (Ellul 1954) and recently from the addiction created by digital artifacts. Awareness is a direct response to these threats. It does not credit the complex machinery behind the scene but it focuses on the scene itself, on ordinary behavior.

In fact, the situation is a little more intricate. Issues like surveillance or addiction can be seen as the intrusion into the ordinary of a non-ordinary and non-acceptable element. The addict of any kind is obviously the victim of some external agency (alcohol, heroin, online pornography, digital "social" networking) but technoethics still suggests taking the problem by the other side. On this other (user) side, the external agency invades the ordinary *of the subject* and this constitutes the problem. The existence of heroin or online pornography is not the problem we should focus on, but rather its intrusion into the everyday life of the addict and the negative consequences of this invasion. The technoethical approach that I put forward intends to assess and to possibly block the process in its ordinary existential reality, as opposed to its external objective reality. There are numerous relevant virtues for this blocking but awareness is typically the basic one. The lack of real awareness that a given ordinary behavior is in fact an addiction seems to be one of the most powerful factors of addiction.

In the scope of this book I take the term "addiction" in its broad acceptation as *the continuation of a behavior despite adverse consequences* and I take tobacco smoking as its paradigm. Tobacco smoking, which is a technology and uses artifacts, is actually the irruption in the most ordinary life (breathing) of an extraordinary technology (inhalation of the combustion products of a plant containing psychotropic molecules). The regular smokers say that they "know" they are "addicts" but that is my point: they need real awareness to dissipate their false consciousness of being a conscientious addict. The angle of attack to restore authentic awareness must be on the level of the ordinary – some practical immediate micro-consequence of smoking, such as being unable to sprint to catch a bus, or spoiling a romantic moment by the urgent need to buy the substance. Extraordinary references, such as the statistics of lung cancer, are apparently less efficient to quit the addiction.

The same pattern of analysis applies to transparency and surveillance in ordinary technology. For the moderately competent human, Edward Snowden's revelations in 2013 about the NSA surveillance program was no news, except for technical details. An elementary digital literacy suffices to be aware of the fact that any digital network communication can be monitored, recorded, stored, and searched. This awareness means that the digital literate would tick "Yes, I know" rather than "No, I didn't know," but it does not mean that in the ordinary course of life one is fully aware of this reality when sending an "explicit" SMS or picture. Therefore, the real concern must be the *ordinary awareness* of surveillance: for instance, the awareness that our mobile phone discloses our position (and the position of others) at any time, or the awareness that any use of a credit card gives away the detailed data of our minute actions and tastes. The relevance of this awareness of ordinary surveillance can be interpreted differently but it must be remembered. For me an experience of this kind was the new "social" function on the Amazon website that offered one day to tell all my "friends" and connections that I just bought an item that could be for different reasons compromising. One click away from social ridicule and grateful it was an opt-in feature.

Once awareness is there, in the ordinary flow of life, as a practice and an effort full of ethical significance, behaviors in the technosphere are not necessarily changed, but their ethical status is changed. I accept the idea of a mindful, perfectly aware and responsible smoker. He or she may be an oncologist for instance. And I still buy a lot of personal things online or with a credit card. But awareness, on the one hand, makes exceptions possible and then the protection of privacy when it is really valued remains possible. On the other hand, awareness can even make surveillance acceptable, even if this is a taboo for some social researchers. People largely know and accept what is disclosed by their credit card or telephone logs. In a sense, surveillance is harmful and intrusive when there is no awareness of it. The awareness of transparency transforms the ethical context of an activity. Empirically, to suppose that complete privacy is a universal aspiration is simply not consistent with the facts in real life: to install almost any application on a smartphone, including gadgets for pure fun, users accept terrifying intrusions into their personal infosphere. They acknowledge deep surveillance by a trivial finger tap on "yes" or "authorize," without which the app would not install. They could live without this app; they can live without strong privacy.

Bruno Latour was right, there are "missing masses" of morality in the contemporary world, and he was right again in situating them in ordinary artifacts such as hotel keys (at that time too bulky to be involuntarily taken away), seat belts in cars, speed bumps in the street, and so on. The next steps in technoethics enlarge and adapt the awareness of these "invisible" agencies. At the same time, we become more and more aware of the moral significance of the ordinary and more and more able to identify moral-laden artifacts. We learn how to detect in them a moral agenda, which we can accept or resist. Silent acceptance can be neither neglected nor mocked since it makes the ordinary of our dealing with the "invisible matter" of morality in the technosphere.

Stupidification and Resistance

What should be resisted, because it directly hinders awareness and then wisdom, is not influence but stupidification. This neologism depicts a type of content and a communication style that are easy to identify once the corresponding sensibility is awakened. Stupidification can take many forms but there is a distinctive feeling when confronted with it: "They think that I am stupid and they try to make me as stupid as possible." If this definition sounds outrageous, may the reader please spend some time listening to TV or radio, or read a popular magazine with its 50-percent advertisement pages, or check a popular news thread.

The loss of content in public communication is one of the real dangers for the contemporary self. An enlightening survey of the phenomenon is given by Richard Lanham in his book on "style and substance in the age of information" (Lanham 2006). Substance has been wiped out by style, by the attractive surface, as a result of the "battle for attention" (Lanham 2006: xi) waged on every media, including personal communication devices. The "economics of attention" deals with the real economic value: attention, which is scarce, and not information, which is overabundant. On a different plane, Hershock (2012: Chapter 5) includes in his critique of universal commodification by the media a strong case against the commodification of attention, saying that it characterizes "the epochal shift from industrial capitalism to mental capitalism and from a material economy to an attention economy" (Hershock 2012: 135).

Movie trailers, magazine covers, catchy book titles, all these callings for attention remain on the surface because they can only reach us in a very short spatio-temporal sequence. Everything (in business) depends on immediate reaction to a stimulus, as it seems. It is safer to rely on the reptilian brain for that purpose. This explains why our streets and our magazines are adorned with images of semi-naked human females or irresistibly appetizing food.

Resisting this systematic stupidification is a noble purpose, which can only be pursued in ordinary life. It requires pragmatic resolution. The first requirement for it is a full awareness of the stupidification process. Please note that the resistance I have in mind necessitates neither an analysis of the machinery behind the scene nor a denunciation of the bad guys behind stupidification. I do not deny the existence of some evil behind the scene, but I focus on the actors that we are on the scene and our capacity to regain an adequate understanding of the scene. Awareness of the stupidification process, hopefully due to educators, can suffice to opt for a resistive attitude. This attitude extends to the integral interface of the self with the infosphere, acting as a spam filter.

Ordinary Wisdom

The contemporary self can consider these awareness practices as active meditation practices. The pragmatic and mundane dimension that is addressed by this sort of wisdom, together with the ambitious objective of self-construction that motivates the practice, are clearly reminiscent of Confucian doctrines with their capacity to consider the secular as sacred and to engage obstinately in the full consciousness of everyday actions (Fingarette 1974). The authentic source of Confucian wisdom is in the ordinary of concrete life, a typical Chinese concern, from its origins, where it was a continued reaction to the abstractness of Hindu philosophical influences (Nakamura 1964: Chapter 15), up to present Neo-Confucian doctrines:

> The proper objects of my moral attention are the actual affairs of my everyday life. I don't need to wait for dramatic opportunities that require courage, compassion and wisdom of heroic proportions.
>
> (Ivanhoe 2000: 66)

Confucian tradition is thus an inspiration for a humble version of self-cultivation through awareness in everyday life and ordinary activities. Its revitalization by Zen (Chan in China) is even more inspiring. Zen develops the capacity to be totally absorbed in the present moment, thing, or activity. "Zen is the *everyday mind*, as was proclaimed by Baso (Mazu, died 788); this *everyday mind* is no more than *sleeping when tired, eating when hungry*" (D.T. Suzuki, Preface to Herrigel 1953: viii). We can add to the list texting when someone is late and looking up a place on Google Maps when one gets lost.

The engagement, significance, and dignity that can be reclaimed by ordinary activities begin in ordinary awareness. Taking a hot shower with a mindful attitude modifies the moral relation to the environmental and economic consequences of the action. On a deeper level it modifies the ethical experience of the self. It can be an authentic experience of life through the experience of a simple technology in an atmosphere of comfort that conveys reassurance and gratefulness and prepares for flourishing and benevolence. The constructive way of conducting technoethics goes from use to practice and then from praxis to ethos. Borgmann sketches an ordinary moral wisdom of this sort:

> We must now try to discover if such centers of orientation can be found in greater proximity and intimacy to the technological everyday life. And I believe they can be found if we follow up the hints that we have gathered from and against Heidegger, the suggestions that focal things seem humble and scattered but attain splendor in technology if we grasp technology properly, and that focal things require a practice for their welfare.
>
> (Borgmann 1984: 200)

The established beliefs state that the most pressing problems are now global and must be addressed on the global level. Then comes the sad observation that no global agent can take action on the global level and then starts the theoretical construction of a global institution that would settle issues of sustainability, peace, trade, energy, climate, data, etc. This global and abstract method has not yet proved it can accomplish something (except for the careers of the experts and bureaucrats involved). A focus on the *ordinary* instances of the same list of questions means a radical relocalization of issues in the concrete existence of the self. I will argue that it has already proved its efficiency and that it can still be greatly augmented.

Awareness and reappropriation of one's ordinary and concrete personal existence engage the third form of power, power over oneself or wisdom, and they disengage the two first forms, instrumental technology and domination (Table 1.3). We could not find global solutions; we need local actions. There is an inconvenient hypothesis to explain our partiality for global and abstract solutions: we secretly know they will never come back to us and will never impose anything harsh upon us. Here again, the ethical move is to replace false consciousness by a more authentic awareness of what happens in the ordinary of one's life.

3.4 Care for Artifacts and Ordinary Attachment

The broadness and depth of our interaction with artifacts invites a realistic technoethics of the ordinary. When looking for relevant data and research in the sociological and psychological literature, one may feel there is a sort of taboo on the emotional and intimate dimension of ordinary life in the technosphere. A phenomenological look is required to see what happens on this scene. The phenomenological effort here will be the effort to bracket the sociological and psychological interpretations that are so pregnant. We are so obsessed by the machinery behind the scene (social and economic determinations or psychological determinations) that we do not concentrate on what happens on the scene of ordinary life.

Future developments in technoethics can follow two promising directions: the ethics of care and the theory of attachment. This section treats only the "object" side of care and attachment in the technosphere; the side of the subject, the self, is treated in Chapter 4 below. As a matter of fact, neither of these sides would exist without its counterpart, as is normal for intentionality in phenomenology. The only strange aspect in this affair is that one of the sides of intentionality is an object and an ordinary object, a kitchen blender or a mountain bike. Do we care for them? Is that properly an attachment?

Bonding with Artifacts

When we feel psychologically, emotionally, and ethically bonded to an artifact, are we only stupid, naive, and victims of the system? Bonding with artifacts is not a moral sin.

Bruno Latour's most stimulating book is perhaps his investigation of Aramis, the aborted French project of a rapid transit train in the 1980s (Latour 1992). The subtitle of this book, "The love of technology," was not ironical, at least no more than Latourian prose in general. Latour's focus was entirely on the design-end. My focus will be on the user-end but our common object is the web of bonds between humans and nonhumans. In this relational network, instrumental rationality is active and always at the forefront of discourse, but a more complex set of motives is necessary for an adequate description of the process. Among these non-technical motives, human passions (courage, resignation, or instances of micro-cowardice) are intermingled with nonhuman sources of agency and counter-agency, such as technical contingencies and the major role of technocracy. France and a lot of countries now have rapid transit trains. They are Aramis' kin. The hybrid network that was described by Latour largely decides their conception and evolution. The time is ripe for considering the hybrid network on the user-end: how users care, or not, for their commuter train; whether and how they are attached to this technological accompaniment of their life.

The same pattern applies to a possible broadening of interesting psychological research, such as Reeves' and Nass' book (1992) in the area of media studies, carrying a subtitle that is not ironical again: "How people treat computers, television and new media like real people and places." This research took place during the last and culminating years of TV dominance in the media and though it may be outdated in its survey of digital communication media it faces with great intellectual courage the ordinary use of the media.

> In short, we have found that individuals' interactions with computers, television, and new media are fundamentally social and natural, just like interactions in real life.
>
> (Reeves and Nass 1992: 5)

> Media are treated politely, they can invade our body space, they can have personalities that match our own, they can be a teammate, and they can elicit gender stereotypes. Media can evoke emotional responses, demand attention, threaten us, influence memories, and change ideas of what is natural. Media are full participants in our social and natural world.
>
> (Reeves and Nass 1992: 251)

This strange psychological relationship has now been extended to the whole infosphere and to large domains of the technosphere where artifacts are smart and kindly interacting with the user. In an immersive video game, the whole sphere of perception is by definition "treated like real people and places." A Web "site" is a real "place" where one "goes" to "do" things. The precautionary quotation marks here do not appear at all in ordinary language. Amazon or Expedia are not taken for human beings, certainly, but they are endowed with a definite personality, which includes moral traits such as a definite degree of sympathy, reliability, honesty, and helpfulness. Every user of similar websites, where people "go" to "do" things on a regular basis, can sketch the psychological profile of the site. And on the designer-end a lot of energy is spent in creating "trust" for the nonhuman object, the website.

Humans obviously interact with digital media rather than just use them. Now, as we have seen, every technology is a media. The whole technosphere is a mediating sphere as we implement everywhere mediating technology more than instrumental technology. Our interaction with it is a "real" interaction. Networks of humans and nonhumans are constantly interacting everywhere. The online tools that support me in writing this text are constantly improved (by humans) and instantly accessible (through nonhumans) in real time as I am writing. The backbone of this constant hybrid interaction is the Internet. It would make no sense to say that we do not "care" for this infosphere and technosphere, even if we are perfectly aware of its not being human.

Caring for Artifacts

Care is a human attitude and feeling based on empathy for particular persons, according to the consensus in psychology and in the ethics of care. But this notion can evolve to cover care as a value that is already accepted in the context of dealing with artifacts (Table 1.1). I am suggesting that the notions of project, purpose, goal, end, and intentionality can be applied to the apparently insignificant facts of life and to the artifacts that support them. This leaves open the possibility for a specific ethical commitment to the ordinary. I am arguing that theories of care and attachment are relevant models to apprehend this domain.

The main change is that mechanical determination from the outside is no longer the privileged explaining pattern for human behavior and for the evolution of hybrid systems. A quasi-mechanical determination from the inside, by mental faculties conceived as mental calculators, would not work either. Complex systemic interaction patterns replace these old paradigms in social and human science, but most of the time they remain on the objective side of reality and do not venture into the subjective side. This subjective side is not exactly blurred; it simply needs a non-mechanical psychology.

There is some uneasiness in losing the illusory causal chain of mental actions and having to manage elusive mental moods like caring and "regarding as important." But the reward is a more precise insight into current situations, says Frankfurt, notably those in which people want or desire things they do not really care about. The novelty is to concentrate not on the artifacts that we want, need, or even use, but on the artifacts that we care about and the artifacts we care for. Our ordinary behavior is more responsive to this grid. "Who needs a Facebook account?" is a poor question compared to: "Who cares about Facebook accounts?". "Why do we use our mobile phone?" is a poor question compared to: "Why do we care for our mobile phone?". Caring for and about a car, computer, clothes, cooking tools, etc. is the ordinary context where values and virtues are mentioned.

The ethical empowerment of the self in the ordinary technosphere can then follow a sort of meditative path: from *awareness* to the realization of *importance*, and from acknowledged importance to *care*. Frankfurt affirms that "the question concerning what is more important is distinguishable from the question concerning what is morally right" (Frankfurt 1988: 82). An ethics of importance is needed in every compartment of moral philosophy but particularly in technoethics. For our purpose it will be centered on the ethical importance of the ordinary. A lot of our ordinary technologies are more important to us than a lot of abstract social values. Probably, from a pragmatic point of view, some are more important than some humans. Our daily behavior has its own ranking of importance and it differs from the established discourse. In this pragmatic scale of values, comfort and facilitation are far above (for instance) social rights in distant countries or environmental consequences. Before assessing this morally questionable trend the philosopher needs to understand it.

The essence of Frankfurt's position is that we should care about what we care about (Frankfurt 1988: 92). Several theoretical advances follow. First, care is a meta-attitude that engages self-care. Second, it calls for the Zen capacity for seeing what is so intimate that it remains unseen: the ordinary technosphere and our ordinary commitments to its artifacts. Third, the meaning of "to care (about/for)" is ambiguous (to attach importance, to feel concern or interest, to feel affection, to like, to look after, to desire, to love, and so forth) but this multiple meaning is a conceptual resource for capturing our relationship to ordinary artifacts and our behavior in ordinary circumstances of the technosphere.

Virtue ethics gives us good reasons to accept a form of "sentimentalism" (Slote 2007: 108). When care is put into the context of sentiment, a declaration of love might follow. Kevin Kelly (2010) confesses that he "loves" the Web. This provocative declaration attracts attention to the unacknowledged affective stories that we live in the technosphere. I would personally prefer moderate expressions of technophilia, and I will reserve the concept of love for human persons. A moderate technophilia can say that we care for the Web. We care for Wikipedia, actively, by using it as a reference and contributing to its articles. There are feelings involved, sympathy and empathy, pleasure in dealing with, joy for its achievements, and sadness in case of harm.

Technophilia can be conceived by analogy with biophilia, the concept offered by E.O. Wilson (1984; see Kelly 2008). Biophilia is an innate tendency in humans to care for life. The broader frame of coevolution encourages symmetry: biophilia and technophilia are the empathic bonds with our two coevolving types of entities. They both express our caring capacities for these entities, our active and affective engagement in coexistence and coevolution with them. Harmony then is not only a consequence of interdependence. It becomes the positive effect of mutual care.

Symmetry and reciprocity prevail in technoethics as in ethics in general. We care for artifacts and artifacts care for us – although I am not certain this is properly perceived in ordinary life. Some examples may help to clarify the obvious fact that the technosphere cares for us. A Google map on my phone, in an unknown city, has been more than once a powerful help to find where I am and find my way. Did it ever prevent me from talking to the locals, to ask for their help, and possibly initiate enriching human contacts? It could be the case, but let us think again. When someone is obviously trying to find one's way with a smartphone, there is no difference with a city map: it suggests to the locals that they may talk to this person and help. In some cases using the Google Map app or just using a smartphone will be considered an incentive to interact since it creates a common culture, a typical "weak tie." Our bonds to familiar technologies can bond us together. Then, shifting to another app, the same phone will help to talk even if no common language is available. In this case, humans care for other humans and it happens in the context of the technosphere where artifacts care for us.

In a normal day at home, a lot of humble domestic appliances care for us. Communication and entertainment devices, all of them, do care for us, especially when we are uncomfortable or incapacitated. And we care so much for our mobile phone that we (should) clean it every day once at least, as we do for (the rest of) our body.

Attaching to Artifacts

Caring for artifacts that themselves care for us does not necessary lead to love but it certainly leads to attachment. Transposed from human psychology to the analysis of human interactions within the technosphere, *attachment theory* allows ordinary technoethics to go deeper into a fascinating dimension. Attachment is coherent with the virtue ethics approach of care and with Frankfurt's vision of care and love: the attachment bond is founded on a caring relationship and usually on a love relationship. Reading the classics in attachment theory while keeping in mind the question of our relationship to artifacts opens promising perspectives at almost every page.

Attachment theory bears on the formation of the bond between the child and the child's caregiver (Bowlby 1969). As such, it is a bond between humans, restrictively. The theory is derived from ethology and it is supported by solid experimental evidence. Its main finding is that the attachment bond is of prime importance in the constitution of the self's psychological structure and integrity. It gives the self a "secure base." The self needs this reassurance background to explore the world and to flourish. Proximity is essential in attachment: the proximity of a precise entity, a human person, the caregiver. Attachment is not dependency (Ainsworth 1972; Bowlby 1977). On the contrary, it provides the secure basis for autonomy, and, more specifically, it gives the capacity to be alone.

The theory investigates attachment "styles" and "internalized working models." Applying these complex notions to artifacts constitutes a vast field of research that exceeds the scope of this book. Some signs of interest for the attachment to objects can be found in academic literature. Yvonne Houy's 2007 article "Living and loving in the metro/electro polis: Understanding the neurobiology of attachments in a society with ubiquitous mobile information and communication technologies" seems promising, but in fact it is all about interpersonal communication mediated by technology. Is there a spell on psychology that limits it to a "humans only" span of attention? Technology is essentially mediation but it is not only mediation between humans. The emotional and ethical dimension in communication technology, for instance, does not concern only the other human(s) involved. It concerns the media. Artifacts matter, emotionally and ethically. Artifacts, including communication artifacts, are substantive nodes in our emotional and ethical networks. Attachment to a transport or a communication device is not only due to its instrumental function in connecting with other human persons. In some cases, the attachment to the device is due to the ironic capacity that it provides for *avoiding* contacts with certain humans (parents or colleagues for instance).

Attachment to artifacts can rely on excellent references but in a strange manner. In one of the founding articles of attachment theory, Harry F. Harlow described the behavior of baby monkeys in comparative experiments with their real mothers and some "surrogate mothers" made of diverse materials. The most successful surrogate mother is described as "a block of wood, covered with sponge rubber, and sheathed in tan cotton terry cloth" (Harlow 1958). Harlow mentions that this situation "is reminiscent of the devotion often exhibited by human infants to their pillows, blankets, and soft, cuddly stuffed toys." He insists on the surrogate mother being an object, a "machine" that dispenses food and warmth. To the present reader, Harlow's irony on the comparative advantages of the artifact-mother sounds outrageous ("The result was a mother, soft, warm, and tender, a mother with infinite patience, a mother available twenty-four hours a day, a mother that never scolded her infant and never struck or bit her baby in anger," Harlow 1958). Strangely enough, attachment theory will remain fascinated by this experimental procedure of the 1950s, which is one of the theory's most frequent references, but the artifact dimension in it remains completely unaddressed (to the best of my knowledge). While exploring affective bonds with artifacts, technoethics in a sense drives back attachment theory to its own origin: bonding with a motherly machine.

Harlow's attachment artifact is analogous to the well-known "transitional object" (Winnicott 1965, 1971). Here we have an attachment bond to a clearly identified artifact. Its importance is strongly connected to the human caregivers but not only to them. Its function is to be transitional between the self and the world. The Teddy, Nanky, cuddly toy, comfort blanket, "binky," or "pacifier" is a part of the self in the world and a part of the world in the self. It embodies mediation in a reassuring form. It is currently remarked that the mobile phone seems to have this function for a number of persons, and this remark is not trivial at all. Toys or bicycles, any artifact can at any age embody the transitional object function for a human person. The capacity to play is essential in the process. A sort of game in the technosphere and with the technosphere is involved in ordinary transitional objects. The transitional object is the first important possession of the self and it is a gift from the caregivers: a smartphone or video game console can easily qualify. Winnicott specifies the role of the transitional object in the capacity to be alone, a capacity that he attributes to the progressive internalization of the reassuring presence of the mother

(Winnicott 1965). The parallel with the progressive internalization of arti-facts is obvious, as is its confirmation from attachment theory.

An ethics of technological mediation is possible from these premises, focused on mediating (transitional) objects and on the attachment for these objects. The secure basis provided by ordinary technology means more than material comfort and social status. It builds the background reassur-ance of the modern self, through its interface with the technosphere and the infosphere.

Combined with attachment to certain humans and sometimes hybrid-ized with them (my grandfather's hammer, for instance), a cohort of ordinary artifacts accompanies our existence. Some of them afford material or digital connection, others provide only symbolic and emotional reassur-ance: the smartphone, the Web, the personal computer, gaming, music or transportation devices, domestic tools and appliances, professional and sports gear, etc. In exceptional cases, the artifact to which humans attach themselves can be extraordinary, a Stradivarius violin or a space station. These cases are already documented and acceptable under the label of exceptionality. What is new with ordinary technoethics is that the objects of attachment are ordinary. The attachment situation is immersed in the everyday course of life. Ordinary matters and here it is constitutive of the value. The standard inter-human attachment is a bond with a typically *ordinary* entity: the parental caregiver. It happens entirely in the most *ordinary* environment, the domestic and intimate sphere of every day and every minute of ordinary life for the baby. Therefore, the link between attachment theory and technoethics is not an analogy but a common exis-tential structure: the ordinary itself, the importance of the ordinary in the constitution of the self. Some sectors of psychology can draw the analogy further: the rituals before falling asleep, which usually involve objects and artifacts, or the attachment of elderly persons to their home environment (so many objects!), an attachment that can be tragically neglected by brutal medicalization.

Attachment to ordinary artifacts has recently received its share of atten-tion in philosophy of technology. Verbeek treats of "attachment between people and things" (Verbeek 2005: 223), and he examines how to establish a bond with an artifact (Verbeek 2011: 101). Almost all the recent studies about mobile phone technology in ordinary life examine some attachment phenomenon or an equivalent notion (Glotz et al. 2005; Goggin 2006; Katz 2008; Ling and Donner 2009; Turkle 2011). Not surprisingly, alas, the mention of this affective bond is most of the time a negative reference and it is interpreted as dependency and addiction. The comforting function of mobile phones in the everyday life of ordinary persons, all over the world and across classes and cultures, remains to be addressed as a pos-itive ethical asset and as an important emotional experience.

References

Ainsworth, Mary. 1972. Attachment and dependency: A comparison. In J.L. Gewirtz (ed.), *Attachment and dependency*, pp. 97–137. Washington, D.C.: V.H. Winston.

Ames, Roger T., and David L. Hall. 2001. *Focusing the familiar: A translation and philosophical interpretation of the Zhongyong*. Honolulu: Hawaii University Press.

Austin, John L. 1962. *How to do things with words*. Oxford: Oxford University Press.

Baker, Lynne Rudder. 2008. A metaphysics of ordinary things and why we need it. *Philosophy* 3: 5–24. doi:10.1017/S0031819108000284.

Bennett, Jane. 2001. *The enchantment of modern life: Attachments, crossings, and ethics*. Princeton, N.J.: Princeton University Press.

Borgmann, Albert. 1984. *Technology and the character of contemporary life: A philosophical inquiry*. Chicago, Ill.: University of Chicago Press.

Borgmann, Albert. 2006. *Real American ethics: Taking responsibility for our country*. Chicago, Ill.: University of Chicago Press.

Bowlby, John. 1969. *Attachment and loss* (3 vols). 1969–1980. London: Tavistock Institute of Human Relations.

Bowlby, John. 1977. The making and breaking of affectional bonds. *British journal of psychiatry* 130: 201–210.

Brey, Philip, Adam Briggle, and Edward Spence (eds). 2012. *The good life in a technological age*. London: Routledge.

Consalvo, Mia, and Charles Ess. 2011. *The handbook of Internet studies*. Hoboken, N.J.: Wiley-Blackwell.

De Certeau, Michel. 1980. *L'invention du quotidien (1). Arts de faire*. Paris: Gallimard. Nouvelle édition (Folio Essais) 1990. (*The practice of everyday life*. Trans. Steven Rendall. Berkeley: University of California Press, 1984).

De Certeau, Michel, Luce Giard, and Pierre Mayol. 1980. *L'invention du quotidien (2). Habiter, cuisiner*. Paris: Gallimard. Nouvelle édition (Folio Essais) 1990. (*The practice of everyday life*, vol. 2. Trans. Timothy J. Tomasik. Minneapolis, Minn.: University of Minnesota Press, 1998).

Ellul, Jacques. 1954. *La technique, ou l'enjeu du siècle*. Paris: A. Colin, repr. Economica, 1990. (*The technological society*. Trans. John Wilkinson. New York: Knopf, 1964).

Fingarette, Herbert. 1974. *Confucius: The secular as sacred*. New York: Harper & Row.

Frankfurt, Harry G. 1988. *The importance of what we care about*. Cambridge: Cambridge University Press.

Gehlen, Arnold. 1957. *Die Seele im technischen Zeitalter*. Hamburg: Rowohlt. (*Man in the age of technology*. Trans. P. Liscomb. New York: Columbia University Press, 1980).

Gibson, James L. 1986. *The ecological approach to visual perception*. Hillsdale, N.J.: Lawrence Erlbaum Associates.

Glotz, Peter, Stefan Bertschi, and Chris Locke (eds). 2005. *Thumb culture: Meaning of mobiles phones for society*. Bielefeld: Transcript.

Goggin, Gerard. 2006. *Cell phone culture: Mobile technology in everyday life*. London: Routledge.

Harlow, Harry Frederick. 1958. The Nature of Love. *American psychologist* 13. http://psychclassics.yorku.ca/Harlow/love.htm.

Hawk, Byron, David M. Rieder, and Oviedo Ollie (eds). 2008. *Small tech: The culture of digital tools*. Minneapolis, Minn.: University of Minnesota Press.

Heidegger, Martin. 1927. *Sein und Zeit*. 13. Auflage, 1976. Tübingen: Niemeyer. (*Being and time*. Trans. J. Macquarrie and E. Robinson. Harper & Row, 1962).

Heidegger, Martin. 1976. Nur noch ein Gott kann uns retten (1966). *Der Spiegel*, May 31, 1976: 193–219.

Herrigel, Eugen. 1953. *Zen in the art of archery*. Trans. R.F.C. Hull. Pantheon; repr. Vintage Spiritual Classics, 1989.

Hershock, Peter D. 2012. *Valuing diversity: Buddhist reflection on realizing a more equitable global future*. Albany, N.Y.: State University of New York Press.

Hershock, Peter D. 2014. *Public Zen, personal Zen: A Buddhist introduction*. Lanham, Md.: Rowman & Littlefield.

Hickman, Larry A. 1988. The phenomenology of the quotidian artefact. In P.T. Durbin (ed.), *Technology and contemporary life*, 161–176. Dordrecht: Reidel.

Hickman, Larry A. 1990. *John Dewey's pragmatic technology*. Bloomington, Ind.: Indiana University Press.

Higgs, Eric, Andrew Light, and David Strong (eds). 2000. *Technology and the good life?* Chicago, Ill.: University of Chicago Press.

Highmore, Ben. 2002. *Everyday life and cultural theory: An introduction*. London; New York: Routledge.

Highmore, Ben (ed.). 2012. *Everyday life: Critical concepts in media and cultural studies* (4 vols). London; New York: Routledge.

Hoggart, Richard. 1957. *The uses of literacy: Aspects of working class life*. London: Chatto and Windus.

Houkes, Wybo, and Pieter Vermaas. 2004. Actions versus functions: A plea for an alternative metaphysics of artifacts. *The Monist* 87(1): 52–71.

Houy, Yvonne. 2007. Living and loving in the metro/electro polis: Understanding the neurobiology of attachments in a society with ubiquitous mobile information and communication technologies. In Sharon Kleinman (ed.), *Displacing place: Mobile communication in the twenty-first century*, pp. 59–76. New York: P. Lang.

Illich, Ivan. 1980. Vernacular values. *CoEvolution quarterly*. www.preservenet. com/theory/Illich/Vernacular.html.

Ivanhoe, Philip J. 2000. *Confucian moral self cultivation*. 2nd edn. Indianapolis, Ind.: Hackett.

Katz, James E. (ed.). 2008. *Handbook of mobile communication studies*. Cambridge, Mass.: MIT Press.

Kelly, Kevin. 2008. Technophilia. *The Technium*, June 8, 2009. http://kk.org/ thetechnium/archives/2009/06/technophilia.php.

Kelly, Kevin. 2010. *What technology wants*. New York: Viking.

Korman, Daniel Z. 2011. Ordinary Objects. *The Stanford encyclopedia of philosophy* (Winter 2011 Edition), Edward N. Zalta (ed.). http://plato.stanford.edu/ archives/win2011/entries/ordinary-objects.

Lanham, Richard A. 2006. *The economics of attention: Style and substance in the age of information*. Chicago, Ill.: University of Chicago Press.

Latour, Bruno. 1992. *Aramis ou l'amour des techniques*. Paris: La Découverte. (*Aramis, or the love of technology*, Cambridge, Mass.: Harvard University Press, 1996).

Lie, Merete, and Knut H. Sorensen (eds). 1996. *Making technology our own? Domesticating technology into everyday life*. Oslo: Scandinavian University Press.

Ling, Richard Seyler. 2004. *The mobile connection: The cell phone's impact on society*. Amsterdam: Elsevier.

Ling, Rich, and Jonathan Donner. 2009. *Mobile communication*. Cambridge: Polity.

Ma, Jianshua, L.T. Yang, B.O. Apduhan, R. Huang, L. Barolli, and M. Takizawa. 2005. Towards a smart world and ubiquitous intelligence: A walkthrough from smart things to smart hyperspaces and UbicKids. *International journal of pervasive computing and communications* 1(1): 53–68. www.yumpu.com/en/document/view/5354361/towards-a-smart-world-and-ubiquitous-intelligence-a-walkthrough-.

Nakamura, Hajime. 1964. *Ways of thinking of Eastern peoples: India, China, Tibet, Japan*. Trans. P.P. Wiener. Honolulu: Hawaii University Press.

Norman, Donald A. 1988. *The psychology of everyday things*. New York: Basic Books.

Piaget, Jean. 1932. *Le jugement moral chez l'enfant*. Paris: PUF.

Puech, Michel. 2013. Ordinary technoethics. *International journal of technoethics* 4(2): 36–45. doi:10.4018/jte.2013070103.

Quine, Willard Van Orman. 1953. *From a logical point of view*. 2nd edn, revised. Cambridge, Mass.: Harvard University Press. 1980.

Quine, Willard Van Orman. 1960. *Word and object*. Cambridge, Mass.: MIT Press.

Reeves, Byron, and Clifford Nass. 1996. *The media equation: How people treat computers, television and new media like real people and places*. Cambridge: Cambridge University Press.

Rorty, Richard. 1979. *Philosophy and the mirror of nature*. Princeton, N.J.: Princeton University Press.

Slote, Michael. 2007. *The ethics of care and empathy*. London: Routledge.

Turkle, Sherry. 2011. *Alone together: Why we expect more from technology and less from each other*. New York: Basic Books.

Verbeek, Peter-Paul. 2005. *What things do: Philosophical reflections on technology, agency, and design*. Trans. R.P. Crease. Pennsylvania: Pennsylvania State University Press.

Verbeek, Peter-Paul. 2011. *Moralizing technology: Understanding and designing the morality of things*. Chicago, Ill.: University of Chicago Press.

Watsuji, Tetsurō. 1996. *Rinrigaku (1937–1949): Ethics in Japan*. Trans. Yamamoto Seisaku, and R. E. Carter. Albany, N.Y.: State University of New York Press.

Wilson, Edward O. 1984. *Biophilia: The human bond with other species*. Cambridge, Mass.: Harvard University Press.

Winnicott, Donald W. 1965. *The maturational processes and the facilitating environment: Studies in the theory of emotional development*. London: Hogarth Press.

Winnicott, Donald W. 1971. *Playing and reality*. London: Routledge.

Zhuang Zi. 475 BCE. *Zhuangzi*. Trans. James Legge. http://ctext.org/zhuangzi.

Zhuang Zi. 1968. *The complete works of Chuang Tzu*. Trans. B. Watson. New York: Columbia University Press.

4　The Self in the Age of Pervasive Technology

Ethics is a philosophy of the self. The eclipse of the self in contemporary philosophy has been an eclipse of ethical concerns, in favor of political concerns, and alas often in favor of ideological prejudices, hard and soft. Participating in this rebirth, the ethics of care and the new approaches to virtue ethics inspire ordinary technoethics in this book.

Questioning technology about its ethical consequences and also about its ethical opportunities means questioning the self in technology. My point is that technology is not intrusive despite being pervasive in the human sphere. Therefore, this ethics of the self in technology will not be primarily oriented toward danger and the protection of the self. It will rather concentrate on the construction of the self. This universal issue seems to be neglected, particularly in its application to the techno-sphere, and even more in its application to the ordinary existential sphere.

4.1　The Return of the Self

The Self as Project in Modernity

The discredit of the subject (the self) in philosophy has the same origin as the promotion of technophobia in philosophy: Heidegger. He begins with an existential analysis where the subject is renamed *Dasein* and where it is (critically) analyzed in the details of daily life, including a famous analysis of tools and technology: the hammer and the car's indicator (Heidegger 1927). Then the famous "Turn" (no pun intended) drives Heidegger to a mystique of Being that disrupts the Western tradition of the conscious subject as the center of philosophy. His critique of technology as a (doomed) destiny is amplified with this turn in his thought (Richardson 1963; Zimmerman 1981).

Foucault begins under discreet Heideggerian influence with a radical deconstruction of the subject as being fabricated by modern institutions of power and knowledge. In his last works, he finds in Hellenistic philosophy a pattern for a resistive construction of the subject, which he connects with the idea of "technique of the self." This evolution from Foucault I to

Foucault II (as there is a Heidegger I and a Heidegger II) shows the way toward a Foucault 2.0, adapted to the present technosphere.

Foucault's last declarations about his own work may embarrass the "textbook Foucault" zealot. My intention is not to address this question directly but rather to investigate the link between Foucault's last constructive formulations concerning the self and the question of ordinary technology. The theme *techniques de soi* ("techniques of the self") was introduced in the third volume of his *History of sexuality*, entitled "Le souci de soi" (Foucault 1984). A better idea of Foucault's stance is given by a series of papers from the same period – "L'herméneutique du sujet," 1982, "L'éthique du souci de soi comme pratique de la liberté," 1984, "Les techniques de soi," 1988 (Foucault 1994: 1172–1184, 1527–1548, 1602–1632). The most interesting documents are the lectures at the Collège de France from the same period (Foucault 2001, 2008, 2009).

Recent philosophy of technology has acknowledged Foucault II and in a certain sense has taken into account the importance of ordinary technology through techniques of the self. The philosophical disruption concerning the self, however, may not be entirely accounted for yet. Verbeek (2011) pleads for a philosophy that would "accompany" technology with a significant reference to Foucault's late considerations about (what I call) the resistive construction of the self: "One becomes a subject not by securing a place outside the reach of power but by shaping one's subjectivity in a critical relation to it" (Verbeek 2011: 73). This is Foucault 2.0 as I see it. Shaping one's life through "self practices" is a form of ascesis that differs from austerity, Verbeek reminds us. This opens the possibility for a construction of the self that goes beyond the shallow moralization of technology envisaged by mainstream philosophy of technology, toward a stronger version where the return of the self is taken in its full disruptive potential. Dorrestjin (2012) concentrates on this question of the self after Foucault. A lot of his tenets on "the design of our lives" pave the way to an ethics of the self in ordinary technology. His understanding of the new "field of technology domestication and embodiment" (Dorrestjin 2012: 114) contributes to a new philosophical context for technoethics.

I argue that a bold focus on the self is of the same importance as the focus on the ordinary. Foucault daringly explained to his fine Parisian audience, with a book of Ancient Greek ethics in his hands, that the reality of philosophy was the work of the self on itself (Foucault 2008: 224, 236). Even more daringly, he explicitly declared that the main questions of philosophy are not politics, not even justice and injustice in the city, but justice and injustice as they are acted by the agency of a self, a subject. For this reason, "the question of philosophy is not the question of politics but it is the question of the subject in politics" (Foucault 2008: 295). Philosophy then entails a "pragmatic of the self" (Foucault 2008: 7) that transfers the question of domination from the *governance of others* to the *governance of the self* (Foucault 2008: 8). "Domination" changes its meaning and its value in the transfer from others to the self – a transition corresponding to the second and third forms of power in Table 1.3.

Foucault's main reference is to the Hellenistic doctrine of *epimeleia heautou* in Greek, *le souci de soi* in French, "care of the self" and "self-care" in English. The notion of *care* in this particular use descends from Heidegger's *Sorge*, the existential preoccupation of the self (*Dasein*) with the world, its engagement with preoccupation, which is usually a derelict form of care. In Foucault's appropriation of this lineage, care for the self belongs to the *technē tou biou*, the techniques of life that prove to be techniques of the self (Foucault 2001). There is a medical (Hippocratic) background in this legacy but Foucault is far from the Platonist metaphors of medical power that traditionally legitimate philosophical power, in the domination sense of power (case 2 in Table 1.3). Care for the self is not a cure of the self; it proceeds neither from the Platonist rejection of the body nor from the religious doctrines of sin. Foucault wants self-care to be *hugieinos*, "life hygiene," and not *therapeuein*, an active "therapy" (Foucault 2008: 10). Life hygiene belongs to the power over oneself (case 3 in Table 1.3); an active therapy would belong to technology (case 1 in Table 1.3). The most important difference is that the notion of life hygiene belongs to the ordinary. Only in the microactions of ordinary life does it make sense to talk of life hygiene. In this sense, the *askēsis* that Foucault puts forward in the practice of the self is an invitation to implement self-care in ordinary life.

Practices of the self as modern *askēsis* are not simply a judicious management of daily affairs. The essence of the self is engaged in these techniques and technologies of the self, which Foucault parallels with the key notion of *Selbstbildung* in German philosophy. The constitution of the self means here the constitution of oneself as a real self (Foucault 2008: 46). This process of constituting oneself as a real self, with all the capacities and dignity of a self, runs through all the references that supported Foucault's meditation in this series of lectures – Marcus Aurelius, Seneca, Epictetus, Plutarch, Epicurus. *Askēsis* in this context does not mean rejecting technology but rather searching for practical wisdom. This practical wisdom remains an open option in the contemporary technosphere and in ordinary technology.

The self is not directly endangered by technology but it may be endangered by the lack of a consistent comprehension of technology. Technophobic tendencies in the humanities on one side and the vain promotion of commodities on the other side both produce this lack of consistent comprehension. Reasserting the self is not rejecting technology but understanding and appropriating technology. History of philosophy has recently tried to popularize spiritual exercises, a tradition "from Socrates to Foucault" (Hadot 2001) that does not boil down to the platitudes of self-help bestsellers. Asian references are a rich resource for this subject and I will use them abundantly – "The basic question of Zen is What is the self?" said Masao Abe (Ives 1992: viii; see also Ames et al. 1994). In constructive visions of modernity, the question of the self and its choices gives birth to an idea of "reflexive" modernity: "The reflexivity of modernity extends into the core of the self. Put in another way, in the context of a post-traditional order, the self becomes a *reflexive project*" (Giddens 1991: 32). Something essential takes place between the self as a project and modernity as a project.

The Ethical Turn

After the empirical turn in the 1980s, philosophy of technology may now be negotiating the curve of an ethical turn. The orientation toward the ordinary is part of this reorientation and it brings with it the question of the self. If humans are destined to dwelling and flourishing in the techno-sphere, the objective predeterminations of this dwelling are not the only subjects that need to be addressed. Dwelling and flourishing are problems of the self. They are modes of being of the self projecting itself into its world. They are styles of the co-construction and coevolution of the self and its world. Ethical life is an integral part of this process.

A simply functional analysis of technology and artifacts reaches its limits when the ethical dimension is reinstated. Functional analysis apparently deals broadly with the user. But "functional" consideration envisages the user-end *from the point of view of the designer-end*. The ethical turn that I suggest affects technology design in a subtle but essential way. It entails a change of perspective and requires a real philosophical effort. Taking the user as a self demands another layer of analysis above (or below) functional analysis.

The return of the subject in the last Foucault is all the more instructive that it comes from the heart of deconstructive philosophy. The contrast of the pages that I have quoted with deconstruction literature and mainstream "Continental" contemporary philosophy is blatant and somewhat uncanny. In his new approach, illuminated by classical wisdom, Foucault sounds optimistic and constructive, at least in the domain of the new field that he discusses: the self. This optimism evokes the elation of a discovery offering unexpected resources.

Does this change mean the return of moral psychology with its characters, mind faculties, and all the scenery of the mind "stage"? This method in philosophy has not been entirely discredited yet but alternative conceptions of moral life emerged with virtue ethics and within philosophy of the self. In the latter domain, Paul Ricœur's narrative theory of the self and self-constitution complements Foucault's notions (Ricœur 1990). The self according to Ricœur is not a fiction produced by external manipulative agents but it is a self-constituting agent. The activity of constituting oneself as a self, through innumerable and intricate frameworks of meaning and their constant reinterpretation, is the constitutive practice of the self. In this approach, morality is embedded in hermeneutic micro-narratives. Ethics is an integral part of the self's life and auto-constitution and therefore ethics is by nature a reflexive agency. Ethics is the reflexive agency that defines the self.

After this ethical turn, the issue concerning values in the technosphere is no longer about a social consensus and its institutional implementation. What is needed is no longer an acceptable ideology; it is a consistent self. "Every good act realizes the selfhood of the agent who performs it; every bad act tends to the lowering or destruction of selfhood" (Dewey and Tufts 1908: 393). Western classics in ethics concur with Eastern wisdom traditions on this point.

Beyond Poiēsis: *Praxis*

Ordinary technologies are examined as practices of the self. This confuses the Aristotelian distinction between *technē* and *praxis* that still influences the common view. Aristotle, in the *Nicomachean Ethics* (Aristotle 350 BCE: I, 1; VI, 4–5) and elsewhere, consistently distinguishes two kinds of human actions: when the end/good/value of the action is external, the action is *poiēsis*, the "production" of something; when the action is (partially at least) the end/good/value in itself then it is *praxis*, a "practice." From this, it follows that *technē* is *poiēsis* and not *praxis*. *Technē* is *poiēsis* with true *logos* says the *Nicomachean Ethics* (VI, 4). For this reason it is entirely different from *phronēsis*, practical wisdom, which is *praxis* (VI, 5).

To reopen the case, technoethics argues that *technē* in Ancient and classical philosophy was restricted to the fabrication of artifacts, whereas contemporary technology is more *used* than fabricated by people. Very few of us build phones and cars but almost all of us use them. The change is not due to technology: in a wooden bed, the Ancients and the Classics see wood and the carpenter's agency but we tend to see primarily the sleeper and the comfort. Our interaction with the technosphere is use more than fabrication. Therefore, our relationship to technology is *praxis* more than *poiēsis*.

The primacy of use implies a change of category in ethics. There is a *praxis* aspect, entailing moral considerations, on the designer-end of technology in the conception and fabrication of artifacts. Engineering ethics is focused on this aspect. But ordinary technoethics addresses the user-end and its specific *praxis*, where moral considerations are different. Here practical agency is different from technical agency; practices of the self engage another project than the instrumental project of *poiēsis*. The self is engaged in the activity of building and modifying *itself*, which is entirely different from the act of building and modifying the world. For this reason "technologies of the self" are not self-contradictory but they must be conceived as *praxis* and not as *poiēsis*, as moral agency and not as technological production.

In the end, not only moral concepts are transformed by this new setting but also general ontology in its distinction between object and subject. There is a candidate to replace both the one-dimensional ontology of the subject (idealism) and of the object (scientific realism): the ontology of the project (Flusser 1994). The constitution of the self (*Menschwerdung*, says Flusser, "becoming human") is conceived as a project that is not purely representational and subjective but essentially engaged in the transformation of the objective world.

In the ordinary, the relevant project is necessarily the individual project, even if is integrated in the project of a small community (couple or family). The transition here will be from an ethics of the subject to an ethics of the project in which the self, no longer an isolated entity, coevolves with a proximal sphere. This proximal sphere of coevolution contains typically hybrid elements, hybrids from the technosphere, from the human and from the environmental sphere. *Project* is a technological notion exactly to the same degree as it is a notion in moral *praxis*. Hence the hybrid character of projects in the technosphere: they are inseparably technological and ethical. For an individual, the project of remaining in the same neighborhood and earning one's living with a job connected to the environment (or to programming, or to social care, etc.) involves a vast array of objects and subjects. Its identity and coherence can still be apprehended as a project: neither as the isolated will of a subject nor as the autonomous machinery of objective circumstances. A global process of coevolution and self-construction is running through this life narrative. What is the self but the sum of its past and present projects, with all their meanings and outcomes?

Heidegger's notion of "project" in *Sein und Zeit* is perhaps underused even in Heideggerian philosophy of technology. It contradicts the simplified version of technology in the instrumentalist paradigm of modernity. A project is not only a process of planning and then executing with instrumental rationality. A project is committed with values, in conceiving it and selecting it, and it meets value issues in the intermediary steps to reach the end.

The project dimension of human existence has never been forgotten but its significance in the ordinary life of the self may have been misjudged. Depriving ordinary lives of project generates existential pathologies in the workplace but also at home and in the life of the city. It is probably not clear enough to everyone that the consistence of a human person goes further than physiological integrity and social integration. Once we are conscious that our dwelling in this world means building – the world and ourselves – it should be clear that we build ourselves and the world in one and the same *praxis*. Existing models of self-construction, from the Ancients to Foucault or Ricœur, in Western as in Eastern traditions, are generally limited to the internal self and occasionally to the intersubjective connection of selves. But now, in the age of pervasive technology, self-construction is inseparable from the construction of proximal, distal, and global technospheres.

Practices of the self are projects in the technosphere. "Technology gives rise not only to disengaged consumption, but also to new possibilities for engagements," writes Verbeek (2005: 190). The project mobilizes the applicative praxis of wisdom (Table 1.4). For instance, resolution can be seen as the composition of courage and autonomy; the acceptability of a project can involve benevolence and harmony. Ordinary technology is often what Verbeek calls mediation in the "mediated engagement." A couple's life project is mediated by SMS and emails but also by domestic arrangements where ordinary technologies and human feelings mix in a strange combination. How to be in love with a smoker or a gamer or a non-driver are examples of continued micro-adjustments in proximal technospheres. The engagement in family life means in certain circumstances an engagement with owning a car and quitting marijuana. Self-construction today takes place in the daily adjustment of such hybrid projects.

We meet Borgmann once again. He pleaded for a redefinition of engagement (Borgmann 1984: 214) and he substantiated it by the idea of "deictic discourse" (Borgmann 1984: 72): the capacity to assert oneself and one's project, as in artistic production. Ironically, the term "engagement," which in French characterized political struggle in its old-fashioned forms (unions and political parties, strike, militancy, revolution), is reborn to name a very different existential and ethical commitment following the return of the self. "If we are to challenge *the rule of technology*, we can do so only through *the practice of engagement*," says Borgmann (1984: 207). But engagement here will refer to Aristotelian *praxis* or Foucault's care of the self.

4.2 Self-care and Consistence

New Beginning in Care

Care is the mode of human openness to the world because this openness is not passive and representational: it is active and technological. From food up to the most abstract culture, we populate our existential world by caring for entities. Things exist "for me" to the extent that I care for them. What I do not care for does not exist for me, in some sense, as long as it does not impose caring for it.

Caring does not only convey the hermeneutic or semantic intentionality of "giving sense to," "acknowledging" something as a distinct entity. Caring as existential openness to things and to the world is an active preoccupation and consists in actions, including decision and volition. The authenticity of preoccupation can be restored on the level of the ordinary – and far from Heidegger's call to authenticity through engagement in an obscure and grandiose destiny, which sounded ominously politicized in the Germany of his time. The ordinary in technoethics is diametrically opposed to the extraordinary destiny in the promise of any ideology. It has its own and Zen-like grandeur and internal truth.

Easier to grasp, the notion of *meta-attitude* perfectly characterizes care. Behind or above first-order attitudes, which can include volitions and intentions, there is a level of second-order attitudes in the human: they are volitions and intentions bearing on first-order attitudes. Classical doctrines of moral character and of virtues deal with these second-order notions (Frankfurt 1988). For instance, when I decide not to answer a given email, my first-order attitude (no answer) is commanded by a second-order attitude, a value or virtue. In that case, it could be the systematic avoidance of conflict, a non-confrontational engagement (a virtue composed of harmony, benevolence, humility and courage, according to Table 1.4). Personal ethics is concerned with second-order attitudes. In a pragmatic version, ethics is mainly concerned with the constant connection and harmonization of second-order attitudes and first-order attitudes. The former guide the latter but the latter constitute the former because second-order attitudes (values and virtues) ethically exist on no other plane but actual behavior.

Care is a typical and comprehensive second-order attitude. It encompasses a vast array of actions and behaviors that have an object and an end other than caring. The secondarity of care explains the reflexivity of care. Self-care then is the second-order project that founds any personal ethical project. This structure is necessary for two reasons: (1) without care, the self is not in charge of itself and there is no ethics; (2) caring for/about anything involves a dimension of self-care – there is necessarily reflexivity in the nature of care. As I argue in this whole chapter, the reason to begin ethics with care and to tie it up strongly in the self is the present situation: the apparent disintegration of the self in multitasking and over-consuming must be driven back to a question of care. The issue with contemporary self is not that it is fragmented and volatile but why it is so. And the answer could be lack of self-care.

I share Slote's central view that care is not simply a special part of morality (Slote 2007): care provides a new foundation for an entire ethics. Technoethics can find a new balance in accepting two of the main consequences that Slote draws from this tenet. The first one is the acknowledgment that care is more foundational than justice. This prepares a de-politicization of the debate in favor of pragmatic ethics. Care is an activity that can be situated in the proximal spheres of the self and does not refer to a global unified system of values as justice did in classical doctrines. The second important point in Slote is the rehabilitation of emotion in the motivational system of virtue ethics. With nuances in the different moments of his work, Slote takes empathy (for other humans) as the key emotion in ethics, following the tradition of British moral empiricism. Here technoethics proceeds boldly further as it includes our relationship to objects in the circle of emotions: not only the other forms of life, which are already commonly accepted as ethically valuable and lovable, but artifacts and in particular ordinary artifacts. Care is then the overall existential intentionality in which everything is considered in its full right.

Foucault's ideas on self-practices and technologies of the self inspired Verbeek's proposal for an ethics of technology where "ascesis consists in developing a conscious and explicit relation to the powers that be, not to master or conquer them, but to govern one's existence with them"

(Verbeek in Brey et al. 2012: 266). The question is "what kind of hybrids we want to be" (Verbeek in Brey et al. 2012: 265), and Verbeek applies it very suggestively to social media. Ascesis is one of the names for self-construction. Verbeek extends its practice to the designer-end, beyond the user-end, arguing that designing technologies can be a self practice (Verbeek in Brey et al. 2012: 272). But in the context of ordinary tech-noethics, where micro-uses and micro-appropriations are the relevant events, co-design in use is typical. The moderately competent computer or smartphone users "customize" their device, that is to say, appropriate it through intensive co-design. This is the reality that I want to consider from the perspective of care. Personalization of one's infosphere and one's prox-imal technosphere is an essential form of care with deep ethical con-sequences – in self-care.

A complex framework of interactions between the kinds of care needs to be explored: care about objects, others, and the self. The usual ethics of care, particularly when concentrated on motherhood and medical ethics, gives a narrow picture of the vast existential range of care. It may prove more adequate to consider care as one of the ethical skills and "know-how" that are acquired in ordinary life through a constant flow of micro-experiences. Dreyfus' phenomenology of everyday "skillful coping" (Dreyfus 2014) is a paradigm to be extended to ordinary ethical coping. Varela (1999) offers a theory of ethical learning in ordinary interactions. The learning process concerning language and logic in Piaget and Quine easily applies to the existential virtuosity that defines ethical care. Through ordinary care received from competent caregivers and through its own continuing practice of care (for others and for oneself), the self evolves a deeply rooted and always improving know-how for care.

From Care to Self-care

The concentration of care on self-care is an essential thesis in Foucault's last philosophy. The meta-attitude of self-care that he called "le souci du souci de soi" (Foucault 2008: 30) contains the same ambivalence in the French word *souci* as in Heidegger's German word *Sorge*. Both convey an idea of being worried and anxious even more than an idea of being simply active in caring for something. The original philosophical notion of care is about being preoccupied more than about being occupied. More exactly, it states that our mode of being occupied is in being preoccupied. Our preoc-cupation invites us to occupation. This existential structure is relevant to the field of technoethics. At first, our preoccupation bears on the necessi-ties of life, food, and shelter, and all the amenities of comfort. But in all this occupation and care, the reflexive nature of care is obvious: self-care is the running preoccupation behind our occupation. You would not care about anything if you did not care for yourself. More philosophically: you would not care authentically for anything if you did not care about caring – and other self-constitutive meta-attitudes.

Power over oneself, wisdom, is not a domination power, neither over oneself nor over others (Table 1.3). This point disengages with the classical and perverse maneuver invented by Plato in the *Alcibiades*: self-care as the preparation of oneself for exerting the supreme power in the city, for exerting the domination form of power, over others. Foucault's series of lectures from 1982 to 1984, about the "gouvernement de soi et des autres" (the government of the self and of others) insisted on the disinterested character of self-governance: it is not a cynical effort to develop the skills for political power. Hellenistic schools of wisdom are not innocent in this perversion of ascesis in Western philosophy: master yourself if you want to master others (and for this purpose only, not as an intrinsic value). The parallel between Marcus Aurelius, Stoic philosopher and emperor, and Epictetus, Stoic philosopher and slave helps elucidate this point. The same Stoic training gives a self the skills to be emperor or slave, indifferently. The problem lies in exploiting this training as a simple means in a strategy for the conquest of political or social power. Politics and greed for power spoil everything, one may think, but it helps to understand how and why: because greed for power spoils self-care and self-construction.

There is a strong connection between care for the ordinary and ordinary self-care. The focus on the ordinary releases us from the perversion of self-care by ideologies of power. When human flourishing is disconnected from domination, self-care can be oriented toward awareness and commitment to the proximal technosphere. I prefer to say that self-care is not in itself a virtue and therefore it does not belong to Table 1.4 ("Wisdom"). One reason is that self-care in its strongest version is arguably identical with wisdom. Another reason is that self-care is certainly the activity that requires the most complete and intense practice of the whole set of virtues. In the list, awareness, autonomy, (internal) harmony, courage, and humility obviously qualify for virtues in the service of self-care. Only one, benevolence, could raise doubt. This doubt is removed by understanding that benevolence for oneself is an indispensable ethical asset. Self-benevolence is an essential virtue for self-care and for ethical behavior as a whole: one of the human persons of value who must benefit from one's benevolence is oneself. It is the necessary condition for a sound taking care of others. Care for others is unsound when it is the outlet for self-hate or self-contempt, as it happens in some religions.

"Reflective life," a life that exemplifies the conception of wisdom in recent virtue ethics publications, implies a deep correlation between care and self-care. Tiberius defines reflective values as being authentic through a specific "endorsement" by the autonomous self (Tiberius 2008: 31). Based on this second-order attitude, endorsement, and based on the "distinction between authentic and inauthentic choices" (Tiberius 2008: 132) that it permits, the lifelong improvement of our current values progressively builds consistence and harmony throughout one's life.

This mode of self-care is perfectly consonant with Frankfurt's (1988) main contribution to ethics: the concept of care founded not on a valuation of the object but on a constitutive experience of the subject. Care is thus disentangled from controversial issues with normative descriptions of the world: care cannot depend on a previous theoretical background consisting in a moral picture of the world. Ethics is directly pragmatic when it is rooted in self-care and even more so if the environment is the ordinary life of the self. The circularity of self-care is not annoying: the self as *project of the self* is not a paradox but the simple structure of reflexivity, which is an integral and definitional part of the notion of self.

Each and every ordinary project is a micro-pattern in the project of the self. The grandiose structures of intentionality that were deployed in abstract logical compositions (Husserl) or gloomy destiny tales (Heidegger) have been redeployed by post-phenomenology within the fabric of daily life. Intentionalities are deeply invested in artifacts and technologies. A new wave in philosophy of technology is now exploring the moral dimension of project and intentionality in the context of our intimate links with artifacts and technologies. A sapiential technoethics intends to go all the way to the ethical intentionality of the self as its own project, as the end of its own practice. Intentionality consists in projection towards the world, but it is essentially a self-constitutive projection. The project for a teenager of assembling an amplifier for a musical instrument, personally conceived and customized out of elementary electronic components, is at the same time a technological project and a self-care project, a self-construction project. There are multiple ways to analyze this instance as a quest for social acceptability through the cult of electronics, or inversely as a resistance attitude against school matters and chores. There is nothing to remove from these interpretations, rather there is something to add: the self projection and the self-constructing value and legitimacy of the project. If the teenager progressively slips into a geek way of life (coding the software for the amplifier all night long for instance), the parental or medical incapacity to understand the self-care value in this activity may lead to a harmful misapprehension of the educative situation.

Self-care and Human Value

Caring constitutes the human in us, therefore there can be no better candidate for being a human value than care and self-care. Classical humanism attributed to humankind a unique intrinsic value, called *dignity* in the Kantian tradition. Dignity relied on humankind's unique capacity to build itself as a value, in Kantian ethics. Humankind inaugurates morality in the world by adding to the realm of nature the dimension of values. This Kantian basis can be challenged but there still remains a cultural consensus on the value created by self-construction. Conscious self-care, beyond sentience, defines humankind and its dignity, probably for a largely universal vision of values. The importance of self-cultivation in Asian thought (Ames et al. 1994) as well and particularly in Neo-Confucianism (Ivanhoe 2000) provides evidence that this represents a universal value.

Self-care has to compromise with care for others and duties toward others. This negotiation normally bears on happiness, one's happiness and the others'. It arbitrates the coexistence of everyone's aspirations to happiness. Kant's doctrine of duties can refine our consensual but blurred ethical concepts on this point. Kant published in 1797 a book of applied ethics, *Metaphysics of morals*, the second part of it being a *Doctrine of virtue*. These texts are coherent with Kant's well-known doctrine of absolute duty and moral law, but in examining the concrete implementation of the moral law in real human life, Kant reaches particularly clear and stimulating formulas. The duties that pertain to virtue are dual, says Kant: my self-improvement ("perfection" in Kant's vocabulary) and the others' happiness (Kant [1797] 1996: Introduction 4, 5, 8). With great insight, Kant insists on the fact that the reverse is not an option: my personal happiness and the others' perfection are not moral duties. Self-improvement is for Kant and for his interpreters (Johnson 2011) something that one owes to oneself, at the core of every moral behavior. It secures the ground for the other moral abilities, those that would impact others. In his vision of human history Kant needs this doctrine of humankind's dignity and intrinsic value. Humankind is characterized by this fate: being in charge of its own construction (Kant 1784). Human persons are accountable and responsible for what they are and the whole ethical edifice is grounded on this humanist acknowledgment. Everything in history, says Kant – and we add: in technology – can and must be conceived as a possible road to self-realization (Kant 1784: Sections 3, 8).

In a very Kantian style, Korsgaard (2009) analyzes human agency on the basis of self-construction and formulates the moral law as a law of consistency. Similarly, in founding ethics as a self-construction process, Tiberius (2008) examines what are the attitudes and meta-attitudes that can and cannot be reflective, ethical, and self-constituent. A differential approach to the notions of *value, meaning, weight*, and *importance* can be found in Nozick (1989: Chapters 16–17), which can help describe specific styles of self-care. To enlarge the picture even more, and to dewesternize it, the best reference is probably the Confucian philosopher Xun Zi whose entire production converges toward a theory of self-completion as the cardinal virtue of the sage (Knoblock 1988).

The rebuke and the accusation is always the same when self-care is assumed as the grounding value in ethics: it is supposed to lead to egoism, selfishness, and individualism, which are allegedly unethical. What is true is that virtue ethics and the ethics of care are able to resist the legacy of obsessional altruism, which might be a subterfuge for domination: the ambivalent desire to do the others' good (by force if necessary) and to provide for their happiness (by being their master) is more than suspect.

Self-care and virtue as I take them are not selfish; a rapid examination of wisdom virtues in Table 1.4 can prove it. *Benevolence, harmony*, and *humility* directly concern others. The remaining applicative virtues are clearly sensitive to others. *Autonomy* and *awareness* are also values when instantiated in others, where they determine their flourishing capacities. The authentic practice of these virtues contributes to an environment of shared awareness and collaborative autonomy. The indirect duty of mutual education expresses the expansion of autonomy and awareness to others. *Courage* is an attitude in front of others but also an attitude for the sake of others.

Another robust rejoinder to the reproach of selfishness comes from the importance of the global for the contemporary self. A new intimacy with the global has materialized in the proximal infosphere and technosphere. The elementary form of this presence of the global is passive: the pervasive presence of all regions of the world in our food, appliances, clothes, music, entertainment, and in every detail of the ordinary environment. Combined with the virtue of awareness and with the engagement in global sustainability through microactions this pervasive presence of the global can initiate an active behavior of the self: a typical non-selfish behavior, which becomes accessible only through the moral maturation of the self.

Moral maturity then should lead to understanding that ethics is not obsession with "others" – *domination* is obsession with others. Since others are selves, the mature moral self understands symmetry between selves as a basic structure in ethics. To care for the others' self-care is legitimate and it is a duty, perfectly consistent with the duty of self-care. Environmental and Eastern ethical perspectives lead to enlarging care to all sorts of sentient entities and eventually to their environment, natural or not, material or not. The paradox that *a deeper self means less ego* is perhaps easily solved through its contraposition: egoism and individualism are consequences of a feeble self. Even without dewesternizing philosophy, it is understandable that a flourishing self is a potential for radiating harmony around the self. "We are imprisoned in our small selves, thinking only on the comfortable conditions for this small self, while we destroy our large self," writes the contemporary Buddhist philosopher Thich Nhat Hanh (Nhat Hanh 2005: 71). Self-care in its highest forms surpasses the conditioned ego and it is immune to individualism.

However, the not-selfish self will not be "altruistic" in the ideological mode transmitted by Abrahamic religions. Resistance to the sacrificial altruism of classical ethics is essential to renew the basis of moral behavior in the twenty-first century. The sound self learns from virtue ethics renewal (Slote 1992) how to be self-assertive without guilt. The applicative virtue of benevolence relies on the resources and potentials of a mature and sound self that is not dragged into the service of others by guilt and self-disparagement. Contributing to a forum or improving a collaborative online article is positive care and self-care. It contributes to the flourishing of the acting self through its taking care of the commons. No weight of duty nor weight of others is felt in this experience, only the elation of self-realization in the collaborative technosphere. The collaborative expansion of the self replaces altruism under duress.

Consistence

The immediate objective of self-care can be called *consistence*. The consistence of the self is its capacity to be and to remain an authentic person in terms of integrity and identity. Thoreau's ideal was a "consistent" man (Thoreau [1845] 2004: 61). Hershock's version of Buddhism for modernity (Hershock 2014: 137) relies on personal consistence, reliance on self-power (*jiriki* in Japanese) rather than on other-power (*tariki*), be it other people or magic formulas (artifacts). His return to fundamentals under the title "reinventing the wheel" describes how an authentic personal biography, a self-endorsed narrative of life, can be regained against what he calls "iconic" scenarios: life models ready-made and sold as commodities along with their material support – wardrobe, house, car, and so on (Hershock 1999: Chapter 10).

I take consistence to mean "holding together" in terms of coherence, and in this sense I privilege pragmatic ethical coherence, namely the coherence between endorsed values and actual behavior. Consistence also implies stability in time; substantial and temporal robustness, but in a dynamic process, in an evolution, a progression, a project.

Self-consistence goes through all the applicative virtues of wisdom (Table 1.4). *Awareness* contributes to self-consistence in providing the reflexivity without which what is built is not a self. *Autonomy* comes next because consistence requires the independence of the self. Autonomy here means the capacity to identify and to assess external influences so that some of them can be actively appropriated by the self as a form of education, not submission. Permanent education and constant attention to self-education are key to consistence as flourishing autonomy. *Courage* will be needed to maintain a perimeter of consistence in the turmoil of interdependence. A perceptive and active capacity for *harmony* will be indispensable for this courage of consistence not to be confrontational. Consistence of the self means independence in interdependence: consistent harmony. *Humility* as a perpetual meditation and practice prevents the consistent self from being a nuisance to others and a self-promoting fallacy. Here the specific values of the ordinary protect from the illusions of moral heroism. *Benevolence* remains a relevant applicative virtue for the consistent self because consistence is connected with the flourishing of all other humans, life forms, and interdependent entities.

We have already stressed the importance of awareness to resist the "stupidificating" momentum of modernity (Section 3.3). The inconsistency in modernity can be linked to inconsistence in the self. The real dereliction of the modern self concerns its own being and it is not imposed on the self by any worldly power. It cannot be remedied by a better design of any sector of the world. Therefore, restoring meaning and substance cannot be the task for deontological or professional codes in engineering or designing; it can never be a top-down "governance" of reality. These levers of public action can only provide the context for self-consistence, which is an essential but not sufficient endeavor. The lack of "content" in the self, the lack of substantial self, as the resource for action, these are enough to explain the circle of stupidification.

Emerson's notion of *self-reliance* describes the capacity of the self to reach authenticity in the context of modern civilization and sociability (Emerson [1841] 1983: 257–282). Thoreau's philosophy treatise *Walden* shows the consistent self in a position not of rejecting modernity, but rather in a lateral option, a step aside (Thoreau [1854] 1985: 321–587). They offer a rich doctrine of self-reappropriation that inherits both ancient and Eastern conceptions of wisdom and then founds them anew in the context of industrial society and the modern state. They perceive that the neglected power, the power over oneself (Table 1.3) is the missing resource for modernity. Self-consistence disrupts the logic of delegation where comfort, material and immaterial, is provided by external dependencies and is passively received in the flow of commodities. Thoreau's intuition was that disruption of this system is simple, for a consistent self.

What is not so simple is to build one's consistence in modernity. Thoreau wants to show that disruptive wisdom starts in the ordinary of the self, in its proximal sphere of action. This lesson has more to do with modernity than is currently understood from Emerson and Thoreau. Contrary to a technophobic reading of their works, they systematically involved technology in the process of authentic self-building – axes and trains, agriculture and the printing press, shoes for walking in the woods, and boats for navigating the Merrimack River.

4.3 Resources of the Self

This section elaborates on two premises: abundance as a fact (in the industrialized world) and abundance as a resource. It forms a hypothesis: a consistent human self can be obtained from this resource and it can address the problems raised by the technosphere, starting with abundance itself.

From Industrial to Digital Modernity: The Digital Self

I take *abundance* in the sense of Galbraith (1958): the contrary of *scarcity*, which is the normal state according to conventional economics. I do not deny the structural injustice that restricts abundance to a limited number of people and countries, and I agree again with Galbraith (2004) on his severe diagnosis. Pragmatically, abundance describes the state where more is available for the 80 percent of a society than the requirements of basic needs. These basic needs can be different according to time, place, and personal options, but I assume they remain in a foreseeable range. *Walden*'s "necessary of life" suggests the headings "food, shelter, clothing and fuel" (Thoreau [1854] 1985: 322). Updating these requirements to a reasonable standard of life in a current industrial society would add a refrigerator, for instance, but not a car. The item-by-item negotiation can be harsh but it will reach a minimum acceptable standard of life that qualifies for abundance. The technosphere provides for material abundance and comfort and the infosphere provides for immaterial abundance, made of information and culture, which is just as important. Let us take the minimum welfare income (social benefits) in European richest states as the threshold of affluence in this restricted sense.

Industrial modernity, thus, gives access to an abundance that reverses the terms of human self-realization. It was not easy to build oneself as a valuable human self in an environment of scarcity and permanent menace. The contemporary technosphere is a new philosophical epoch: it is not easy to build oneself as a valuable human self in an environment of affluence and permanent distraction – but for new and unexpected reasons. This paradox of the affordances of affluence is central to technoethics. The first philosophers of the self in modernity, Rousseau in the eighteenth century, Emerson and Thoreau in the nineteenth century, offered answers to this problem, all of them relying on the existential resources of the self and on education (self-education in priority).

The overall pattern of coevolution includes circles of coproduction between technological cultures and technological selves (Margolis 1989). This coproduction is not mechanical determination: we are shaped by the technologies that we shape. The awareness that we shape these technologies belongs to the *reflexivity* characterizing a self and leading to wisdom. When I buy my first smartphone, I know that some aspects of my life will be changed by the availability of emails, the Web, GPS, and so on. The decision to buy it depends not so much on the financial trade-off; it concerns the existential trade-off. When money matters more, in decisions about owning a car for instance, the essential trade-off is still existential: I am choosing my self, in a sense, not only the brand or the color of the car. The processed food package that I am buying (or not) is an act of self-construction. The question of buying it or not is a micro-decision in self-building. In front of both stairs and an escalator, I choose my self in terms of health and psychological disposition, in terms of symbols as well, and I contribute to the survival of one in two artifacts populations (stairs or escalators). The coproduction of selves and mundane technosphere entities must be understood on the micro-level of ordinary life.

The infosphere accentuates and systematizes this type of situation. Because a text message or a click can deeply affect human lives, the density of interactions within the infosphere imposes on us a permanent existential responsibility. Always potentially online and with a full power of decision, the modern self is summoned to sustain its self "in real time." The digital self, then, is not a fixed and predetermined self like the *subject* in classical Western philosophy. It can only be a constantly self-forming and self-transforming self in a permanent search of its nature and coherence. The modern self is arguably the process of self-construction and not the static entity resulting from it. Real *agency* in coevolution as in self-evolution is substantial to being a self in the technosphere. I can choose my self only if I choose myself.

There is then a difference between the *proximal* and the *intimate* for the digital self. Intimate means a constituent of the self; proximal means a resource for the self. The mobile phone or laptop is proximal, some data in it are intimate. Replacing the device is a change that needs a couple of hours or days to be integrated, whereas changing certain data (alteration or loss of intimate data) infringes on the self.

Technosphere and infosphere are spheres of the self. The human self, except for an extremely small elite, was restricted to one sphere only in every historic period, except for the last centuries or decades. The life of women in the industrialized world was long restricted to the famous "three Ks" in the German expression (*Kinder, Küche, Kirche*: "children, kitchen, church"), each of them narrowly channeled. When the vast majority of the population was working in agriculture and living just above the survival level, we know from historians that their horizon was strictly limited and uniform. This narrow existential sphere still is the rule in some parts of the world.

Spheres of the self in industrialized modernity are infinitely diverse and mutable. This is an anthropologically disruptive condition. The variety and affluence of material and cultural spheres is apparent at the global scale, with digital media now giving access to immense resources in a fraction of a second. On the Web we can "surf" from human sphere to human sphere in a way that was not even imaginable before.

In this new environment, the proximal technosphere and infosphere of the modern self are typically informal. We do not know exactly what we possess nor what we know. We often need to check and search the house, cellar, and garage for an item that may well have been thrown out or lost or put away in the most improbable place. A *radical simplicity* experience can start with a total emptying of the house on the lawn in front of it. The resulting heap embodies the nonsense of our acquisition culture. The items for sale in a supermarket or a mall are hundreds of thousands – only computers know, and imperfectly. What is available for sale or for access on the Internet, nobody knows. This is the informal technosphere and the informal infosphere of abundance. The self deploys and incessantly reconfigures its proximal spheres, drawing on the resources of global spheres. What is available in the global spheres (Web or supermarket) largely depends in return on the billions of proximal spheres formed by the billions of selves. Ordinary technoethics tries to intervene in this cycle at the point where the self constitutes its own proximal spheres.

The incessant reconfiguration of one's ordinary technosphere and infosphere forms the flow of ordinary life. It can be Facebook updating or tweeting, but more ordinarily it will be adding a new contact in the phone or mail directory, adding or deleting a bookmark in the Web browser, getting micro-information that induces the tentative adoption of a particular aliment, and so on, minute after minute in ordinary life.

Spheres of the self might be supposed constitutionally "open," as in "open source" for software. But human nature has it that ideological enclosures try to prevail everywhere, it seems. Everywhere enclosing spheres appear with the project of locking up people and resources in a circle of control and domination. It has been dubbed the risk of cultural "siloing" on the Web, silos being the closed infospheres. Some already exist, for instance islamosphere or nazisphere, the spheres of anorexic addicts and maybe the obsessive fans of some artist.

The cases of selective infospheres and technospheres are instructive. The most interesting practice is certainly the Amish way of life right in the center of the most industrialized consumer society, the USA (Sclove 1995). Radical simplicity engagement or a selective education by family or institution, through cultural or religious influences, can produce harder or softer versions of selective infospheres and technospheres. In all these cases, supply is limited for the constitution of the self's proximal sphere. The existential situation of the self is not modified in its nature but in its parameters.

Constructive Attitudes

A constructive attitude is at work for most of us in our real life, personal and professional, or at least we must charitably suppose so. Academic approaches to social and individual phenomena have long been obsessed by deconstruction. One can argue that it does not necessarily mean narrow-minded destruction (or stagnation) but it certainly does not always mean open-minded construction. The return of the self in philosophy gives a timely opportunity to test an ambitious constructive attitude. Investigating the multiple ways in which the modern self is dissolved, fragmented, and split remains a necessary duty of human and social sciences, but it cannot be their sole contribution. Constructive human and social sciences can contribute to the constitution of the self in the technosphere and not only prophesy its dissolution by the technosphere. The "enjoy and criticize" attitude (brave researchers crucifying marketed technology, but on their brand new MacIntosh) is twice non-constructive: on the one hand it does not change the real state of affairs and on the other hand it gives (to the brave researchers) a false consciousness of critical awareness.

An entirely opposite logic of constitution can be drawn from *askēsis* as self-constitution in Foucault (Foucault 2008: 305). Coming from the ancient world, this ethic fits perfectly into the digital world: the good life is self constitution, self education, self valuation, in an uninterrupted effort of self-awareness where life is learning, training, building oneself – a "formation" in Foucault's French (Foucault 2008: 421), "la formation de soi comme un soi."

In technoethics, figures of the self in technology are linked to an infinite variety of ordinary flourishing circumstances. An emblematic case is the difference made by open source or collaborative software, for instance

Firefox (Web browser), Thunderbird (email) or LibreOffice (office software). Personalization is a core feature in each of these programs, through collaborative add-ons and options that go deep into the software functions. The user contributes to the software in more than one sense: through extensively customizing one's own tool and through possibly participating in collaborative incremental improvement. This collaborative participation may require no programming: just posting the description of a bug helps, just commenting on the features or add-ons helps, just sharing one's "macros" or "skins" (program's appearance) helps. These constructive behaviors build resource spheres that interconnect in the infosphere and that supply selves with tools, inspirations, comfort, and reassurance. These are actions of care and most of the time they also mean something in terms of self-care, self-construction.

The capacity of the digital sphere for substantiating a self allows fake identities, inevitably. The Hindu notion of *avatar*, the temporary embodiment of a god in a creature, was in use in video games and it was afterward popularized by online social networking. *Second Life* became famous as a video game and a social network for avatars, fake selves. Formerly, some writers could have absorbing imaginative lives and they would "upload" significant aspects of their life into a literary avatar – Marcel Proust's "narrator" for instance. In any epoch, some mentally perturbed subjects have been diagnosed with split-personality disorder, more or less severe and crippling. But now everyone can take a pseudonym and interact with other humans as an entirely different person, without any limit or consequence, apparently. The problem with an avatar's creation and inhabitation is precisely how easy it has become and how this has changed the rules of identity in the infosphere – to be or not to be a Marcel Proust or a psychotic cyber-troll.

Double life and multiple identities can be a game and a game only: the studies about online multiple selves tend to normalize playing with multiple selves, particularly for teenagers who are still in the process of constructing their identity through trial and error (Turkle 1995; Markham 1998; Joinson 2003). Virtual selves belong to the sphere of the self as projection, tentative self, or just fun. As in any other circumstance of life (offline), experimenting with a virtual identity online can be a constructive experience. It can be a resource for flourishing as a richer self. It can also be a deconstructive experience for the self, a bad experience that will remain as a trauma. The same self-construction logic works for traditional cultural supports such as fairy tales, books, and films. The difference with digital immersive technologies is in their existential efficiency, so to speak. Immersive engagement is a fundamental feature in recent video games, some of them being called "first person" games for a good reason.

With such possible flourishing resources that may turn into danger for personal integrity, the philosophical question bears on *means and ends*. Once one's ends are established, that is to say managing importance and care, the self can flourish even in possibly dangerous encounters.

Experiences of all sorts, in this way, will end up in wisdom, hopefully. But when means are confused with ends, or when real ends are obscure to the acting self, the same powers that work for self-construction are at work for self-destruction. Sport is a paradigm for this ambiguous process. Physical activity must be included in a personal life plan, for physiological and psychological reasons that are well known. For some, sport can be a personal engagement toward performance, at least during a phase of their life. Then it will imply diet control and a set of competitive quantification in several aspects of life. A lot of devices make this control process considerably easier today. Why not then a performance enhancement drug? The issue is not about it being legal or not (entire sports are contaminated by doping, even if there is a media blackout about it for obvious business reasons). The issue is about means and ends. The authentic end is self-flourishing; it is not record time, speed, wins, or points. Constructive resources turn destructive when the end is lost.

A philosophical test of these concepts is the case of the perfect drug. Let us imagine the invention of a substance that produces maximum pleasure and satisfaction, a paroxystic physical and mental elation (sexual orgasm is clearly the model), a substance without any physical damage or physiological addiction, legal, and inexpensive. My question is: what would be the reason not to try it? Or the reason not to take it on a regular basis? Or the philosophical reason not to consider it a fascinating opportunity for human flourishing? An ethics that has no answer to this challenge may be inadequate for the immediate future and actually for the present. Consistence is the answer suggested by the technoethics that I put forward.

Flow and Access

We do not have the perfect drug, or not yet, but we have the TV, which is close to it, and we have video games and online activities that can be addictive. The question is to implement in the ordinary infosphere an ethics of self-construction and flourishing. To sort out the multiple features of these experiences I will use the distinction between *flow* and *access*. Perceiving this difference generates a quasi-physical perception of what happens in the infosphere and in some parts of the technosphere.

A *flow* media transmits information in one way only, from emitter to receiver, with an imposed timing and without any communication feedback from the receiver. A flow media is typically destined to a passive receiver in an informational asymmetric pattern where the receiver is submitted to the media.

An *access* media gives access to something else than the media itself – which is not as trivial as it may sound at first. TV usually leads nowhere but to more TV; it is a largely self-contained flow and not just a self-promotional media. Mass media work as a largely auto-referential flow. On the contrary, the Web can lead to real things and places, to actions and real persons. On access media, the user, who is no longer a receiver and who is no longer passive, retrieves the information. This information is pulled by the person and not pushed to the person – *push/pull* is a basic distinction in media theory. Active users act with their own timing (moment and rhythm). In the most important access media (the Internet), users are allowed to change the content they access or to publish their own content. The Web is close to the perfect access media. Books are typical access media, even in their paper form, and libraries of books, CDs, DVDs, and so on are physical access media of great social importance.

The current primacy of flow media over access media explains much of the present situation in the global infosphere. In particular, the present rule of flow media explains the neglect and loss of content that characterize our cultural environment.

It makes all the difference for the self: being exposed to a flow versus being granted access to a resource. Only the latter promotes flourishing through the personal infosphere. The ordinary self looks up something, finds it and uses it, in a constructive project. The digital literate completes such micro-cycles all the time, there may be dozens of them in writing a single page for school or office, there are hundreds of these micro-retrieving cycles before finding a new job or after a medical diagnosis. Formerly, the greater part of one's knowledge came from what has been taught to the person and passively learned by her. Personal search for knowledge was not a daily activity except for some professionals. Today, retrieving information through a digital media is a daily activity for the majority of people in industrialized countries. Our individual knowledge stock is more and more full of personally retrieved information. The same situation prevails in acquiring skills and of course in the existential experiences online. Flourishing occurs more and more in the context of a voluntary activity and less and less as the result of passive learning.

Autonomy in self-flourishing is not only a value; it is the practice of a virtue. In this perspective, almost everything in our schooling system must be questioned. This reconsideration concerns the absurd idea of an initial period of training that leads to a degree that determines one's entire life course. Continuing education and permanent training could prevail instead. The obsolete practice of a master talking to a constrained audience locked in a room, which is the practice of education in many places, becomes a real hurdle on the way to culture and self-flourishing. School is still essentially a flow media and it must transform itself into an access media. It means de-schooling education and de-schooling society – another claim made by Illich (1971) decades ago.

On the whole, transforming flow media into access media is an ongoing task of prime importance for the flourishing opportunities of human persons. The Web, emails, and social conversation online is more and more enriched by contents "hacked" from TV or flow media, re-published online or sent as a link, for access. The flow is captured and redisplayed in access mode.

Flow media pollute the global infosphere because they insufficiently participate in the cycles of self-construction that make the contemporary cultural ecosystem. They run for their own sake and their own benefit in a closed trade circle. In the opposite direction, access media are under the threat of being transformed into flow media. From AOL to Microsoft, numerous media corporations have attempted to transform the Web into a subscription flow media. Flow vs. access is a battle in the cultural war for sustainability and it is an essential battle because it bears on the resources for self-building, the foundation of ethical life: the resources of the self.

Self-governance and Generativity

The recent notion of *governance* was coined to bypass the discredited notion of government. Governing oneself can be called self-governance, to bypass political connotations. Self-construction depends on available resources, which abound in the infosphere and technosphere, but it depends essentially on governance and self-governance. To become a self and oneself requires governing oneself into a self. The applicative virtue of *autonomy* is here at work and an ethics of self-governance clearly falls within the project of wisdom and existential consistence.

More and more smartphone applications present themselves as a direct support to self-governance. This function must be distinguished from the "quantified self" movement, which was typical of the first epoch in electronic self-surveillance. The "apps" that I have in mind can be very simple, as simple as buzzing once a day to remind the person that she must take her contraceptive pill – micro-dosed pills tolerate no oversight. The app can follow one's weight daily and connect the phone directly to the scales. In e-health, more sophisticated apps are already available, with various bio-data acquisition interfaces combined with the autonomous capacity to analyze them and alert the owner or the appropriate medical service. Personal agendas and task managers proliferate in the infosphere. They can track any task, backward and forward, including the most intimate projects. But their added value, compared to traditional methods such as paper and ink, lies not in high-tech extraordinary features. I remember monitoring with obsessive precision and implacable enforcement my working hours for competitive exams at a time where paper and ink were the sole ordinary self-management tools. Now I can do the same on the computer or smartphone screen, using a dedicated software or simple universal tools. I would not even say that the digital device helps or increases the performance of my personal workflow management: my laptop and smartphone being always with me for other reasons they simply "attracted" my self-governance practices (a case of convergence in technological coevolution). The shopping list or workflow control chart is always available in the computer or the "cloud," or even better they are in the pocket (in the phone). Self-governance, thus, is an activity and an attitude that uses extraordinary tools while they are available, but it can do without them and it has its value in itself. When it is induced by the hype around a fashionable app, it might collapse rapidly and remain as just one more episode in stupidification. Real change means deeply affecting ordinary practices and most of the time affecting ordinary self-practices. Self-incentives and self-monitoring do not require a smartphone but they can use it. The functional advantages due to the high-tech device do not need to be huge or even to be real because their real functional superiority is their integration into the proximal existential sphere, which is partially now a proximal infosphere.

How far can this assisted self-governance go? Borgmann has perceived the danger of *reverse adaptation*, "the adjustment of human ends to match the character of the available means" (Borgmann 1984: 60). This danger is frequently mentioned in relation to ordinary technologies: people are said to telephone only because they own a brand new telephone. Can one be said to cook just for the pleasure of using one's new kitchen or to run just to enjoy one's new shoes? Surely, but where is the problem? Even under a condescending description (by a "social scientist" bravely fighting capitalism), cooking and running remain valuable activities, particularly if these practices belong to consistent self-care. Devices and commodities are powerful incentives: the problem is when, why, and how to enroll them as self-incentives. Self-governance is more and more the management of self-incentives which we accept to insert, or not, in the proximal technosphere. A real virtuosity in ordinary technoethics is needed here.

The cover image of Sherry Turkle's book *Alone together* (Turkle 2011) demonstrates how the common assumptions about technology can conceal technoethical values. This picture shows four people walking while watching their mobile phone – or simply absorbed by their phone screen. But we do not know what is actually displayed on these screens. We can arrogantly suppose it is bullshit; all the scenery is devised to induce this assumption. We must charitably suppose, on the contrary, that these people are planning a meeting in a nearby café, where they will spend hours chatting and flirting. Now let us imagine the banal book cover with the four of them sitting together and having fun. How would we know that the meeting was arranged from scratch an hour ago thanks to mobile phones? Even *reverse adaptation* can be a flourishing opportunity for human persons. Smartphones have invented new social opportunities.

Zittrain's book on the opportunities and threats for the Internet defines a suggestive notion: *generativity*. It is "a system's capacity to produce unanticipated change through unfiltered contributions from broad and varied audiences" (Zittrain 2009: 70). The Internet is obviously the paradigmatic generative system. I would say that a technological system *allows* or *affords* generativity, in a range starting at "tolerates generativity" and ending at "boosts generativity" or even "works only from generativity." Real generativity comes from the agency of human selves who are engaged in personal and collective projects largely exceeding the domain of the considered technology. Following the argument in this chapter, a subsequent notion would be: *self-generativity*, defined as the self's capacity to build itself from generative elements in the technosphere and infosphere. In the infosphere, the term *selfware* is already in use and I wish it a bright future.

4.4 The Good Life in Ordinary Technology

The central notion in the technoethics that I put forward is wisdom as a virtue and not as a normative concept of the good. Nevertheless, the question of the good life can be addressed from this point of view. The main inflexion that I want to give to the current ethics of the good life is the focus on the ordinary technosphere and infosphere.

Refocusing the Good Life on the Ordinary Technosphere

Charles Taylors's influential book about the "sources of the self" stresses the link between "selfhood and the good," two "inextricably intertwined themes" (Taylor 1989: 3). Taylor also emphasized the importance of ordinary life. A lot of his intuitions and case analyses about "the affirmation of ordinary life" (Taylor 1989: Part III) are consonant with the technoethics presented here, except for one reservation about Taylors's religious option. Ordinary life and its autonomy are referred to by Taylor in an argument to restore *transcendence* in our secular age. I will rather take *immanence* as the acceptable "source" of the self and of the good in technoethics, following my previous argument about human self-constructed moral dignity. I reach opposite conclusions, naturally. We need no transcendent supplement for founding the ethics of modernity. On the contrary, we need a deeper understanding of the immanent resources in the technosphere and in the entire modern environment. The grand question of the good has been transposed into the ordinary but it was for deprecating the ordinary and for restoring transcendence.

Anthony Giddens expressed the transition to the ordinary differently (1991). While questioning self-identity in modernity, Giddens depicts modernity as it existed just before the digital revolution. Giddens' vision of the modern self supports my hypothesis about ordinary technoethics more than Taylor's stance. He advocates a reflexive awareness and the importance of lifestyle choices on a typical ordinary level:

> Modernity is a post-traditional order, in which the question, "How shall I live?" has to be answered in day-to-day decisions about how to behave, what to wear and what to eat – and many other things – as well as interpreted within the temporal unfolding of self-identity.
>
> (Giddens 1991: 14)

He detects the resistive component of ordinary reappropriation: "Even the most oppressed of individuals – perhaps in some ways particularly the most oppressed – react creatively and interpretatively to processes of commodification which impinge on their lives" (Giddens 1991: 199). His final recommendation is called "life politics" and it sounds typically depoliticized or post-political, pleading for "a remoralising of social life" and "a renewed sensitivity to questions that the institutions of modernity systematically dissolve" (Giddens 1991: 224).

Existential awareness is the relevant philosophical concern. The existential reappropriation of technology in the ordinary, aiming at the good life, differs from previous reappropriation doctrines, which were more or less ideologically motivated. Ideology was previously counted as an advantage for motivation and group cohesion. A new system of values would systematically prefer non-confrontation: the good life cannot be gained by winning in a fight; it has not to be gained against opponents. Good things are not systematically issued from battles, where we, the good ones, vanquish and eventually destroy the others, the evil ones. History has long been written and taught this way but now we can decide that confrontation needs no longer to be systematically kindled. The good life aspires to be non-confrontational and it may be the case that modernity affords it.

A non-confrontational attitude toward technology favors neither massive adoption nor massive rejection but selection and appropriation. Therefore, the question is: how? Why a smartphone but not Facebook, for instance (or the opposite choice)? Why an SMS to this person now and not a Skype call? This is a question of *judgment*, in the technical sense of the word, the Kantian sense (not a demonstrated knowledge) or the Aristotelian sense (practical wisdom which computes the indefinitely diverse parameters of a given concrete situation). The skills and training for this sort of judgment are dramatically underestimated and hardly ever mentioned. My point is that they can shape a particularly relevant art of living for our age. The aim is not an "alternative technology" as it was in the 1970s but an alternative good life in technology.

Meta-literacy, Meta-empowerment

The reflexive dimension of ethics responds to the modern abundance in resources and solicitations. To put it in a nutshell: it requires a strong self to be a human person in the present technosphere. It seems natural to favor acting on the technosphere itself, on its backstage mechanisms, in order to instate the self in a condition of autonomy and flourishing. I advocate acting directly on the self instead.

Empowerment is a pivotal value in technoethics and a meta-attitude necessary to wisdom. Real empowerment is neither directly the information provided by Google or Wikipedia nor the material capacities offered by a modern car or credit card. Empowerment is in the second-order capacity to invest this first-order affordance with a second-order dedication to self-construction. Any online or material resource can be seen as an ambiguous challenge for flourishing and the good life. This challenge requires agency. The online or material resource will never be a direct "good" to be stored and secured in one's existential stock, so to say. The particular agency or meta-agency that is here needed is the meta-culture of modern culture: a meta-culture of access management for the self, so that it can thrive on the affordances provided by the technosphere and infosphere. Specific resources for well-being in digital technology abound (Amichai-Hamburger 2009) but they are not "ready to use" commodities.

The interesting notion of "digital literacy" (Prensky 2001; Ulmer 2003; Berry 2012) must expand in the way the infosphere has expanded to embrace the whole technosphere in its ordinary pervasiveness. The modern self is solicited many times in a single day in occasions that call for tech-noethical literacy. The ultimate recourse for response to these micro-solicitations must be internalized in the self, it must be a product of self-care and actually it is the practice of self-care in itself.

Concerning "smart" artifacts, this literacy makes the difference between *being an object* in the networks of techno-structures and *being the subject* of a human life. A finger tap on the phone screen makes me visible, invisible, or more or less easily visible, on the radar of data greedy corporations who spy on the micro-data in my ordinary life – it can also be anxious parents or unethical employers. "Dear customer, we will let you know when it is ready with a text message, may I have your phone number?": the acceptance of a service makes us a target for unsolicited solicitations (called spam) which are currently spoiling the commons of digital space. In all these instances, the self must be able to invent micro-resistance strategies. I call this skill "outsmarting the smart" (ad-blocking software is doing the job quite well, for instance).

The abundance of resources for the good life in modernity will produce at its best an epoch-making pluralism in the forms of life chosen by self-reliant selves for their own flourishing. There is a humanist version of digital ethics (Ess 2009: 183–199) and a demanding theory of how to sustain pluralism in it (Hershock 2012). The richness of the present and future cultural world resides in the pluralistic set of values and practices that can now be chosen in the immense repertoire of human flourishing capacities. Access to this existential repertoire is more and more independent of physical and social place. Canada for instance is a good place to flourish as a Buddhist. The French countryside is a good place to flourish as a programmer. Even an unemployed person in an industrialized country (with reasonable social welfare programs) can access information and culture virtually without limits – there are cultural limits, and this section is dealing with them, but few material limits.

Borgmann discusses the question of life orientation in modernity with a reference to Kant's view on intellectual and ethical orientation (Borgmann 1984: 218). Borgmann deplores the modern situation where this fundamental choice is under "the rule of technology": "The rule of technology is not the reign of a substantive force people would bear with resentment or resistance. Rather technology is the rule today in constituting the inconspicuous pattern by which we normally orient ourselves" (Borgmann 1984:

105). But this rule is at the same time a resource and it is a resource for resistance. Borgmann himself detects the exact locus for this resistive reappropriation of the technosphere: ordinary life and its micro-engagements. The good life, he writes, "is a matter of daily practice, of acquiring moral skills and habits, of keeping them sharp, and of exercising them regularly" (Borgmann 2006: 29).

Ordinary Values for the Good Life in the Technosphere

Multiple instances of a determining link with the ordinary are suggested by the literature on the good life in a technological world (Higgs et al. 2000; Brey et al. 2012; Verbeek 2013). The provisional table of apparent values in technology (Table 1.1) can now be assessed and modified in order to converge with the good life model that emerges from the previous and present chapters.

In the first row of ordinary values, (1a) *comfort (well-being and reassurance)* and (1b) *facilitation* are wholly confirmed and substantiated as values for the good life in the technosphere. They are obvious resources for the flourishing self. There is a change in the valuation of comfort, from a consumer value to a self-reassurance background. This is a paradigmatic change brought by ordinary technoethics. In this perspective, (1c) *production and consumption (economic growth)* is no longer acceptable either as a primary end or as the prescribed means for the good life. The apparent value of prosperity, at least under its production and consumption form, may be the most harmful dogma of modernity.

The "background values" of our provisional table, (2a) *power* and (2b) *control* as self-sufficient values, have been dismissed by the analysis of power (Table 1.3), which criticized both dominations: the domination of things via technology and the domination of people via political or institutional power. A major disruption, an ethical one, is introduced as soon as we stop using power and control as ends and self-sufficient values. A great number of decisions and choices, in personal as in institutional life, would be modified by this change. When considered in the abstract, the majority of enlightened contemporaries would agree that a radical downsizing of the importance of power and control would certainly bring us on the way to the good life. But when facing real choices, in dramatic as in ordinary situations, our spontaneous option remains power and control. The deconstruction of this prejudice belongs to the incessant personal practices of wisdom that I detail in Chapter 6.

The emergent (row 3) and potential (row 4) values in Table 1.1 were evoked on a regular basis in the argument of this book or will be active in the following chapters. The project of ordinary technoethics accentuates the set of constructive values that emerges from the technosphere and infosphere. It thus candidly validates the impression that in a lot of places (geographical and social), life in the present world is better, it is more of a good life, than in previous forms of industrial or pre-industrial civilization.

The technophobic prejudice that systematically counts technological innovation as a hindrance to human flourishing is not only contrary to common sense and common experience; it proves to be contrary to a constructive ethics.

Health as Self-care in Technological Affluence

Health is everywhere the most wished-for blessing for a human person. It comes now before riches and victory in battle. One of the reasons for this appraisal is that illness represents the most violent irruption of fate in our controlled lives. Medical technology is available to repair, but not always, and in the end it will fail. This common vision relies on a shallow ideology of technology as a "quick fix." It leads to the absurd representation of death as one fatal disease which still evades medical treatment. The philosophy and ethics of health can offer better approaches but health remains the basic condition that the self hopes for and benevolently wishes to others. In the modern mentality, it is not exactly under the form of hope and wish; it is a project, an undertaking. We know that we are actively in charge of our health and this is a paradigmatic sector of self-care in the technosphere.

The distinction between extraordinary and ordinary technology is particularly relevant in this matter. The promise of technology cannot be to ensure survival and perfect health by the means of maintenance and repair devices for the human body. This would be a total domination of nature in mode (1) of power (Table 1.3): a total control of things and nature by technology. It is in fact the project of transhumanism, and I have argued against this option (Section 2.1) in favor of a coevolutionary logic for which harmony is a greater value than total control and domination. The good life does not require permanent optimal physiological parameters, not even normal parameters.

The question of health therefore has to be framed in terms of ordinary self-care instead of techno-medical miracle. Health is maintained and repaired by ordinary self-care, concerning food and physical activity, well-balanced working time and sleeping time, and a safe management of toxic and addictive substances and behaviors. Life hygiene belongs to the ordinary lifestyle and it matters. An important part of this self-care exceeds the proximal sphere, in particular the pollutants in the air, water and food, and all the environmental (natural and technological) parameters that are impacted by collective and global behaviors. For this reason, the external conditions of the good life require taking in charge the commons and taking care of the global commons. This aspect will be addressed below in Chapter 5.

Self-realization

Self-realization, one of the expressions of flourishing in the good life, is inclusive of other humans and of the environment – all kinds of environments, particularly natural and social environments. Arne Naess based his proposal for modern wisdom on the essential connection between *self-realization*, central in his ethics, and the capacity of *wonder* in nature, the other focal point in his ethics (Naess 1989, 2005). Self-realization occurs in the interdependence with other life forms. The access to a deep and authentic experience of nature, for this reason, deserves to become a fundamental human right. Restoring the link with nature means an emotional and symbolic experience of the totality of being and it can be interpreted as the genuine endorsement of coevolution.

The link between self-realization and the social refers to another network of interdependence where wonder is no longer the emotional link. Nevertheless, the authentic perception of interdependence between humans must be sensible, something felt and not just known. The self can engage social issues from a perception of interdependence with a greater authenticity than when its engagement proceeds from a theory of justice or academic utilitarianism. Ordinary technology, and particularly digital communication technologies, offers multiple opportunities for this patient rebuilding of the social, self by self, through multiple connections. Recent studies demonstrate how the mobile phone in various circumstances can be a determining resource for people most in need – all sorts of outcasts, social, economical, cultural, and political (Katz 2008). Even when it results in massive social movements, this social action of a new kind starts from an individual device, which initiates the first action: "capacitating" the self.

Heikkerö suggests that Borgmann's concern about modernity can be summarized in the word "unfocused" or "disintegrated" (Heikkerö 2012: 108). An article by Briggle and Mitcham (2009) coined the notion of "disembedded" for the modern self. Refocusing on the self is a response to deconstructive trends in modernity. It offers an original strategy for restoring focus, integration, and embedding. Heikkerö illustrates what he calls "restoring a focus to life" by the importance of specific focal things and practices in Japan (Heikkerö 2012: Chapter 4): *kendō* (a martial art linked to the Way of the Sword), judo and aikido, *chadō* (the tea ceremony, which is literally the Way of Tea), and *ikebana* (the art of floral arrangement, which is *kadō*, the Way of Flowers). Things and practices are involved in these cultural forms but they are typically understood as *ways*. The way in question is the way to wisdom. Conceived with pluralism and tolerance, it includes but is not restricted to Buddhist illumination (Okakura 1906). "In fact, nearly anything can become a focal practice for a Zen practitioner," concludes Heikkerö (2012: 123). Without being a Zen practitioner, the modern self can take charge of its ordinary experience with the resolution of refocusing his life and self.

Ironically, the self is endangered not by contemporary technology but by contemporary philosophy. The post-modern deconstruction of the subject and the reign of objectivation in social science (to vainly mimic hard science) left the post-modern self apparently helpless in a massively hostile technosphere. The fact that the technophobic orientation of intellectual media and institutions does not match the common experience of ordinary technology is not a problem: it fuels the self-assigned mission of the elite. My intention is a climate change in the assessment of technology potentials. The title of this chapter was certainly understood in its first reading with a connotation of compassion for the poor self in the predicament of technology. This title should be read again with a connotation of optimism and even enthusiasm for the challenge of being a self in a resourceful technosphere.

References

Ames, Roger T., W. Dissanayake, and T. P. Kasulis. 1994. *Self as person in Asian theory and practice*. Albany, N.Y.: State University of New York Press.

Amichai-Hamburger, Yair (ed.). 2009. *Technology and psychological well-being*. Cambridge: Cambridge University Press.

Aristotle. 350 BCE. *Nichomachean ethics*. Trans. W.D. Ross (1925). http://classics.mit.edu/Aristotle/nicomachaen.html.

Berry, David M. (ed.). 2012. *Life in code and software: Mediated life in a complex computational ecology*. Open Humanities Press. www.livingbooksaboutlife.org/books/Life_in_Code_and_Software.

Borgmann, Albert. 1984. *Technology and the character of contemporary life: A philosophical inquiry*. Chicago, Ill.: University of Chicago Press.

Borgmann, Albert. 2006. *Real American ethics: Taking responsibility for our country*. Chicago, Ill.: University of Chicago Press.

Brey, Philip, Adam Briggle, and Edward Spence (eds). 2012. *The good life in a technological age*. London: Routledge.

Briggle, Adam, and Carl Mitcham. 2009. Embedding and networking: Conceptualizing experience in a technosociety. *Technology in society* 31(4): 374–383. http://dx.doi.org/10.1016/j.techsoc.2009.10.001.

Dewey, John, and James Hayden Tufts. 1908. *Ethics*. New York: Henry Holt.

Dorrestijn, Steven. 2012. *The design of our own lives: Technical mediation and subjectivation after Foucault*. PhD Thesis. University of Twente. www.stevendorrestijn.nl/downloads/Dorrestijn_Design_our_own_lives.pdf.

Dreyfus, Hubert L. 2014. *Skillful coping: Essays on the phenomenology of everyday perception and action*. Oxford: Oxford University Press.

Emerson, Ralph Waldo. 1983. *Essays and lectures*, ed. J. Porte. New York: The Library of America.

Ess, Charles. 2009. *Digital media ethics*. Cambridge: Polity Press.

Flusser, Vilém. 1994. *Vom Subjekt zum Projekt. Menschwerdung*. Bensheim und Düsseldorf: Bollmann.

Foucault, Michel. 1984. *Histoire de la sexualité, vol. 3. Le souci de soi*. Paris: Gallimard. (*The history of sexuality, vol. 3: The care of the self*. Trans. R. Hurley. Vintage, 1988).

Foucault, Michel. 1994. *Dits et écrits I, 1945–1975*. Paris: Gallimard (Quarto, 2001).

Foucault, Michel. 2001. *L'herméneutique du sujet. Cours au Collège de France (1981–1982)*. Paris: Gallimard/Seuil. (*The hermeneutics of the subject: Lectures at the Collège de France, 1981–1982*. New York: Picador, 2006).

Foucault, Michel. 2008. *Le gouvernement de soi et des autres. Cours au Collège de France (1982–1983)*. Paris: Gallimard/Seuil. (*The government of self and others: Lectures at the Collège de France, 1982–1983*. New York: Picador/Palgrave Macmillan, 2011).

Foucault, Michel. 2009. *Le courage de la vérité. Le gouvernement de soi et des autres II. Cours au Collège de France (1984)*. Paris: Gallimard/Seuil. (*The courage of truth (The government of self and others II): Lectures at the Collège de France, 1983–1984*. New York: Picador/Palgrave Macmillan, 2012).

Frankfurt, Harry G. 1988. *The importance of what we care about*. Cambridge: Cambridge University Press.

Galbraith, John Kenneth. 1958. *The affluent society*. London: H. Hamilton.

Galbraith, John Kenneth. 2004. *The economics of innocent fraud: Truth for our time*. Boston, Mass.: Houghton Mifflin.

Giddens, Anthony. 1991. *Modernity and self-identity: Self and society in the Late Modern Age*. Cambridge: Polity Press.

Hadot, Pierre. 2001. *La philosophie comme manière de vivre. Entretiens avec Jeannie Carlier et Arnold I. Davidson*. Paris: Albin Michel (*Philosophy as a way of life: Spiritual exercises from Socrates to Foucault*. Hoboken, N.J.: Wiley-Blackwell, 1995).

Heidegger, Martin. 1927. *Sein und Zeit*. 13. Auflage. Tübingen: Niemeyer, 1976. (*Being and time*. Trans. J. Macquarrie and E. Robinson. London: Harper & Row, 1962).

Heikkerö, Topi. 2012. *Ethics in technology: A philosophical study*. Lanham, Md.: Lexington Books.

Hershock, Peter D. 1999. *Reinventing the wheel: A Buddhist response to the information age*. Albany, N.Y.: State University of New York Press.

Hershock, Peter D. 2012. *Valuing diversity: Buddhist reflection on realizing a more equitable global future*. Albany, N.Y.: State University of New York Press.

Hershock, Peter D. 2014. *Public Zen, personal Zen: A Buddhist introduction*. Lanham, Md.: Rowman & Littlefield.

Higgs, Eric, Andrew Light, and David Strong (eds). 2000. *Technology and the good life?* Chicago, Ill.: University of Chicago Press.

Illich, Ivan. 1971. *Deschooling society*. New York: Harper & Row. www.preservenet.com/theory/Illich/Deschooling/intro.html.

Ivanhoe, Philip J. 2000. *Confucian moral self cultivation*. 2nd edn. Indianapolis, Ind.: Hackett.

Ives, Christopher. 1992. *Zen awakening and society*. Honolulu: Hawaii University Press.

Johnson, Robert N. 2011. *Self-improvement: An essay in Kantian ethics*. Oxford: Oxford University Press.

Joinson, Adam N. 2003. *Understanding the psychology of Internet behaviour: Virtual worlds, real lives*. New York: Macmillan.

Kant, Immanuel. 1784. *Idea for a universal history from a cosmopolitan point of view*. Trans. Lewis White Beck. In Immanuel Kant, *On history*. Indianapolis, Ind.: Bobbs-Merrill, 1963. www.marxists.org/reference/subject/ethics/kant/universal-history.htm.

Kant, Immanuel. 1996. *The metaphysics of morals* (1797). Trans. Mary J. Gregor. Cambridge: Cambridge University Press.

Katz, James E. (ed.). 2008. *Handbook of mobile communication studies.* Cambridge, Mass.: MIT Press.

Knoblock, John. 1988. *Xunzi: A translation and study of the complete works* (3 vols), 1988–1994. Redwood City, Calif.: Stanford University Press.

Korsgaard, Christine M. 2009. *Self-constitution: Agency, identity, and integrity.* Oxford: Oxford University Press.

Margolis, Joseph. 1989. The technological self. In E.F. Byrne and J. Pitt, (eds), *Technological transformation: Contextual and conceptual implications,* pp. 1–15. Dordrecht: Kluwer (*Philosophy and technology,* vol. 5).

Markham, Annette N. 1998. *Life on line: Researching real experience in virtual space.* Walnut Creek, Calif.: Altamira.

Naess, Arne. 1989. *Ecology, community and lifestyle.* Cambridge: Cambridge University Press.

Naess, Arne. 2005. *Selected works, vol. X: Deep ecology of wisdom – Explorations in unities of nature and cultures.* Dordrecht: Springer.

Nhat Hahn, Thich. 2005. *Being peace* (new edition). Berkeley: Parallax Press.

Nozick, Robert. 1989. *The examined life: Philosophical meditations.* New York: Simon & Schuster.

Okakura, Kakuzō. 1906. *The book of tea.* Dreamsmyth edition, 2011. www.gutenberg.org/ebooks/769.

Prensky, Marc. 2001. Digital natives, digital immigrants. *On the horizon 9(5).* Bingley, UK: MCB University Press. www.marcprensky.com/writing/Prensky%20-%20Digital%20Natives,%20Digital%20Immigrants%20-%20Part1.pdf.

Richardson, William J. 1963. *Heidegger: Through phenomenology to thought.* 3rd edn, 1974. La Haye: Nijhoff.

Ricœur, Paul. 1990. *Soi-même comme un autre.* Paris: Seuil (*Oneself as another.* Trans. K. Blamey. Chicago, Ill.: Chicago University Press, 1992).

Sclove, Richard E. 1995. *Democracy and technology.* New York; London: Guilford Press.

Slote, Michael. 1992. *From morality to virtue.* Oxford: Oxford University Press.

Slote, Michael. 2007. *The ethics of care and empathy.* London: Routledge.

Taylor, Charles. 1989. *Sources of the self: The making of the modern identity.* Cambridge, Mass.: Harvard University Press.

Thoreau, Henry David. 1985. *Walden or life in the woods* [1854]. In *A week on the Concord and Merrimack rivers; Walden or life in the woods; The Maine woods; Cape Cod.* New York: The Library of America.

Thoreau, Henry David. 2004. *The Higher Law. Thoreau on civil disobedience and reform.* Ed. W. Glick. (H.D. Thoreau, *The writings*). Princeton, N.J.: Princeton University Press.

Tiberius, Valerie. 2008. *The reflective life: Living wisely within our limits.* Oxford: Oxford University Press.

Turkle, Sherry. 1995. *Life on the screen: Identity in the age of the Internet.* New York: Simon & Schuster.

Turkle, Sherry. 2011. *Alone together: Why we expect more from technology and less from each other.* New York: Basic Books.

Ulmer, Gregory. 2003. *Internet invention: From literacy to electracy.* New York: Longman.

Varela, Francisco J. 1999. *Ethical know-how: Action, wisdom, and cognition.* Redwood City, Calif.: Stanford University Press.

Verbeek, Peter-Paul. 2005. *What things do: Philosophical reflections on technology, agency, and design.* Trans. R.P. Crease. Pennsylvania: Pennsylvania State University Press.

Verbeek, Peter-Paul. 2011. *Moralizing technology: Understanding and designing the morality of things.* Chicago, Ill.: University of Chicago Press.

Verbeek, Peter-Paul. 2013. Resistance is futile: Toward a non-modern democratization· of technology. *Techné* 17(1): 72–92. http://dx.doi.org/10.5840/techne 20131715.

Zimmerman, Michael E. 1981. *Eclipse of the self: The development of Heidegger's concept of authenticity.* Athens, Ohio: Ohio University Press.

Zittrain, Jonathan. 2009. *The future of the Internet: And how to stop it.* New Haven, Conn.: Yale University Press.

5 Deep Sustainability and Personal Microactions

My intention is not to overlook the collective, the social, and the global but to reconfigure them. The main reorientation lies in the banal analogy of bottom-up orientation replacing top-down orientation. Ordinary technoethics gives a solid base for non-trivial implementations of the bottom-up aspirations that emerge in our contemporary set of values. Global issues can be revisited and ethically overhauled by ordinary technoethics. The global agency that we are apparently unable to establish by the way of institutional action may be discreetly available in the disseminated ordinary. A *deep* sustainability, based on technoethics, could be the notion we need to reshape the collective, the social, and the global in accordance with our best hopes.

5.1 Extended Meanings for Sustainability

The ethical implications of sustainability and globalization often remain implicit or underestimated. On the contrary, I will argue that ethical sustainability is the imperative we have to measure up to and that ethical globalization is the challenge for our times.

Ethical Sustainability

Ethical sustainability is not a strange concept because the concept of sustainability refers to values – or it should. The purely functional approach to sustainability addresses the continuing performance of a system, to be maintained across changing environmental and internal factors. A purely utilitarian framework is involved. More precisely, it is a purely functional efficiency pattern, where no value is mentioned except the "neutral" value of efficiency itself. The canonic "Brundtland report" did not criticize this meaning for sustainability in the world economy and ecosystem and for this reason it was ambivalent (Brundtland 1987). Governments and corporations rushed into the possibility of taking sustainability as a conservative ideology: maintain the same economic and political system by adapting to environmental and societal constraints (Puech 2010). The ambivalence leads to nonsense.

Sustainability entails (functional) ecosystemic stability and autonomy. Beyond this, the idea of *acceptability* in "sustainable" is of prime importance today. The Oxford Online Dictionary clearly gives "able to be upheld or defended" as the second sense of sustainable, on par with "able to be maintained at a certain rate or level." The French original word "soutenir" means to support, hold, maintain in the material, assertive, and moral senses of these English verbs. The semantic drift toward ethical meaning and away from the functional meaning may be a sign of the times. It instantiates the notional shift that I suggest for global issues: the primacy of *ethical* sustainability. Economic sustainability and political sustainability more and more depend on it and must be reconsidered from its perspective.

Global ethical sustainability has held the attention of leading figures in contemporary ethics. In his 2002 book about "the ethics of globalization," titled *One World*, Peter Singer affirms that this problem has not yet been correctly addressed by other branches of human thinking and activity. When Singer quotes Thomas Friedman saying that the most basic truth about globalization is that "no one is in charge" (Singer 2002: 1), I suggest adding the statement: "everyone is in charge," and then connecting both statements like this: "no one is in charge (politically, institutionally), therefore everyone is in charge (ethically)."

The awareness of an ethical "charge" concerning global sustainability is more and more manifest in societal conversation, even if the exact nature of this feeling is not yet settled. Rifkin (2009) widely propagated an idea of emotional globalization as *empathy*, framed by the "global consciousness" that emerges from the contemporary infosphere and the contemporary cultural moment in history. The new bond for the global is both emotional and ethical under this description, like the notion of empathy itself. This thesis challenges institutional and intellectual establishment: how could an informal and ordinary psychological mood provide a basis for real social construction? Institutions claim monopoly for addressing social issues. The discontent with global official policies and the ordinary feelings of empathy for all sorts of victims may change the agenda and impose a new deal.

Rifkin is not specific on the sources of empathy in the self. In our tentative grid of the applicative virtues of wisdom (Table 1.4), *empathy* would be composed of awareness, benevolence, and harmony. A particular form of empathy for the global pertains to ordinary technoethics if we add humility (no plan to save the planet as a Hollywoodian hero), autonomy (no intention to submit to an ideology or an institution before taking action), and of course courage, without which empathy is nothing but a passive feeling. An active empathy is then a comprehensive expression of the applicative virtues of wisdom. Compassion is fundamental in Buddhism; empathetic care for the global characterizes Chinese thought, both Confucian and Daoist. Rifkin's conclusion is surprisingly somber:

> The problem is that the same communications technology revolution that is paving the way toward global consciousness has a dark side

that could derail the journey and sidetrack the Internet generation into a dead-end corridor of rampant narcissism, endless voyeurism, and overwhelming ennui.

(Rifkin 2009: 554)

The ethical and subjective dimension appears here only in the negative. Empathy is from the beginning an ethical emotion of the self. Despite its current digital mediation and globalized extension, it must be addressed on the level of the self, as a resource and an agency of the self.

An intriguing new form of ethical sustainability is emerging in academic and public policy circles: responsibility toward future generations. Hans Jonas' initial argument extends our responsibility as far as our power extends. At the same time he asks us to acknowledge that our power now reaches future human generations. This course of thinking is perfectly valid, but it deserves a more constructive and tolerant elaboration than the technophobic development by Jonas himself (Section 2.4). Dieter Birnbacher (1988) offers a more promising model for generational sustainability without an ideological agenda. He states very clearly that sustainability cannot mean the sacrifice of present happiness in favor of some hypothetical happiness in the future (mine or others'), the motto of many an ideology. His call is ethical and not emotional. A remarkable effort in Birnbacher's research on the ethical extension of our consciousness to the future is his adapting the question of norms in the case of incomplete knowledge concerning the consequences of our present action in the future (Birnbacher 1988: Chapter 4 and 6). This approach is a way to evade scientism and self-assured futurology.

Ethical sustainability means *deep* sustainability because it reaches the fundamental issue of interdependence and addresses it with an overall strategy: harmony. This specific level of thought was clear to the Stoics whose moral holism was of cosmic dimension. For them, harmony with *logos* and nature required the same effort, and it was essential for happiness and virtue (Cooper 2012: 151). The Stoic ethics of contributing to the "world-spirit" or world-mind (the *logos*) as the way to wisdom is analogous on numerous points to Hershock's modern Buddhism and its contributive virtuosity in karmic encounters (Section 6.3, 6.4). This shows that Western conceptual limitations contribute to the present predicament but they can be circumvented by the past of philosophy or by the East.

Global issues are oriented toward a possible deep sustainability when global issues are identified as a new kind of problem that transcends the capacity of the technostructure and of all existing institutions (Hershock 2012: 4–8). The Buddhist perspective on the unity of all beings and Hershock's own vision of a specific collaborative dynamic for flourishing out of diversity and interdependence lead to a very particular holism. The "primacy of relationality" (Hershock 2012: 11) initiates a deep ontological change: the individual is no longer a separate rational (calculating) entity, as it is typically in the dominating utilitarian discourse about global issues.

The ambition to offer "an ethics of interdependence" (Hershock 2012: 16) begins with a shift from "deliverable goods and exploitable resources" to "relational domains" (Hershock 2012: 229). On a deeper level it relies on "a new structure of feeling" (Hershock 2012: 241) where the contribution to harmony is intimately felt and not concluded:

> We are in need not just of an ethicization of the public sphere, but an approach to ethical engagement that is freed from the axiomatic, modern conviction that we are essentially independent individuals who enter rationally into social contracts with one another for the primary purpose of maximally securing our own self-interest.
>
> (Hershock 2012: 159)

A New Political Sustainability

The most interesting extension of sustainability is linked to the new forms of social action that periodically feed the news since the 1999 street demonstrations against the WTO summit in Seattle. The sociologist of the networked society, Manuel Castells, has kept an eye on these "indignation" movements, from his homeland's (Spain) *Indignados* to the various *Occupy* and 99% movements throughout the world (Castells 2012). If we take the global infosphere as a key structure in the global political sphere, then the online activist movement Anonymous and websites similar to Wikileaks belong to the same category of social action. Castells insists on their *hybrid* nature: online and in the streets of the real world. He relates this feature to the essential property of being *inclusive* networks.

The global era is networked and networking. Political protest under its new form is only one manifestation of this nature. Hybrid networks of individuals and communities, infosphere sites and physical places, local and global protestations: everywhere the hybrid and networked nature of these movements mark them as a new form of global politics, typically blurring the limits of institutional structures and inventing post-political action. Some of these movements address relatively local problems, for instance the so-called "Arab springs," notably the Tunisian and Egyptian revolutions in 2011 (Castells 2012: 22–31, 52–109), but the vast majority of them embrace global issues – even when it is on a national basis such as the resistance to the world financial system in Iceland in 2008 (Castells 2012: 31–44). Indignados, Occupy and Anonymous introduce themselves as a new agency taking over old institutions with the intention to finally address global challenges. These challenges are very aptly construed by these new movements as being inseparably political, economical, and environmental. These emerging actors personify a form of care for global sustainability, which arguably is the main issue for modernity. The "indignation" trigger is typically an individual reaction. It may also be the appropriate motive for political or social engagement, in the name of decency more than in the name of justice.

The "anonymous" cultural value, beyond its picturesque folklore, conveys the important idea of a grassroots movement where the individual is the *ordinary* individual: anonymous does not only mean *hidden*, it means *ordinary*, anyone amongst the many. A common strategy for escaping surveillance is to transmit via anyone's computer, using massively collaborative software to route the data, preferably with the consent of the intermediary self who "anonymously" joins the resistance in the infosphere. The "occupy" appellation recalls the fundamental question of dwelling (Section 2.4). It poses it appropriately in terms of a reaction to the uninhabitable conditions imposed on a majority of people by financial structures (and their political subsidiaries). The new capability to be local in the global, through bottom-up networks for sustainability, starts with the project and the capacity to authentically dwell somewhere. Reappropriating the local commons gives access to the resources that emerge when global tools connect self-reliant persons and communities. Resilient networks are built from self-reliant persons.

These new social movements are criticized because they lead nowhere in terms of institutional change. However, I agree with Castells that these movements do not intend institutional change; not because it is not wished but because the political and institutional level is no longer relevant for post-political movements. This interpretation of what is important in social movements is a major fact for political science. I consider it from only one particular viewpoint: the emergence of ordinary life as the relevant level in politics. The confrontation is clearly between the 99 percent of ordinary people, now empowered (virtually) by the ordinary technosphere and infosphere, and on the other side the 1 percent elite. They do not live in the same places, they do not go to the same hospitals, and they do not send their children to the same schools. The 1 percent does not live in the ordinary world (Stiglitz 2012).

Common people are not in charge of the commons; the ordinary is not officially connected to the extraordinary global challenges of the technosphere and ecosphere. My hypothesis is that the commons, including the intellectual and information commons (Tavani 2004: 252–260), can be sustainable if their fate is considered from a broader and deeper ethical view, which implies ordinary people and is rooted in ordinary life.

The first appearance of the commons as an issue in modern times was due to a paper in *Science* by Garrett Hardin (1968). It was an alert concerning the "tragedy of the commons," a general phenomenon that consists in the destruction of common spaces (pasture, river, etc.) or devices (a well, a mill) because every individual overuses the commons and nobody is in charge of maintaining them. Hardin was aiming at one single problem, population growth, and he had in mind a debatable solution, "mutual coercion." His article belongs to an uninterrupted series of alerts concerning the unsustainability of our overuse of global resources. The Club of Rome report (Meadows et al. 1972) and the Brundtland report (Brundtland 1987) were landmarks in this global alert, which has now reached every citizen on the planet. Officially, the global is more and more understood as global *commons* for which humankind as a whole is in charge. Except for some rogue states or firms, every community on the planet is supposed to be engaged in the "ecological transition." But it can be argued that the incumbent institutions, public and private, are structurally unable to manage the transition to sustainability and are just pretending to (Puech 2010).

Elinor Ostrom's research on the successful methods for governing the commons suggests that ordinary technoethics can contribute to and in fact be inspirational for new kinds of institutions, devised for tackling sustainability issues. The reconsideration suggested by Ostrom (1990) is radical: what has to be governed and what governing means both require a severe *aggiornamento*. She surveys how local communities everywhere in the world manage "common pool resources" (water, forest, fish etc.). Some of them are remarkably successful in creating and maintaining local bottom-up institutions that are eventually assimilated into existing institutional structures (Ostrom 1990: 1). She offers a table of the design features that characterize the long-enduring local institutions of this type (Ostrom 1990: 90, 180). She specifies the methods of constant self-improvement that made them durable organizations. She concludes wisely: "Learning is an incremental, self-transforming process" (Ostrom 1990: 190), which is true for both humans selves and communities.

Acceptability: Societal Sustainability

A new type of concern can be coined *societal sustainability*. "Societal" is a convenient notion to free social questions from the particular kind of politicization that comes from the nineteenth century (trade unions and social class struggle) and to directly address the relationships between human persons and communities. The societal level is obviously impacted by economic and institutional domination, but the ethical approach to sustainability aims to be a pragmatically anterior consideration.

The prominent effect of the growing importance of societal sustainability is certainly the question of *acceptability*. Technological developments in electronics, such as RFID chips, or in biotechnology, such as GMO in general, raise societal concerns about their safety and also about their control. A trade company easily understands that the acceptability issue must be included in the design project, as early as possible, and that it will be crucial to the implementation process of an innovative product. Seen from the user-end, the question of acceptability is the question of *acceptance*, which engages specific applicative virtues in technoethics: awareness, harmony, autonomy, courage, benevolence, and many composed sensibilities and practices involved in micro-coevolution processes. Non-acceptance can take the form of street demonstrations and militancy, an organized boycott for instance. But *ordinary non-acceptance* is more common and more powerful. It remains on the level of self-agency. Just not buying or not using can make a difference, without any confrontation, through ordinary microactions.

Acceptability issues are then representative of a new style and a new substance in social life: an informal conversation is running online and offline across lines that are more intricate than before. Marketing departments know that explicit discourse, in the media or in answers to inquiries, and even on collaborative forums, is only the emerged part of the iceberg. The main part is invisible because it is disseminated in ordinary actions.

Let us examine a salient case of acceptability in order to flesh out this idea of societal sustainability and its link to ordinary technoethics: the acceptability of risk and in particular of nuclear risk. Social sciences provide rich and applicable patterns of analysis for risk acceptability (Douglas 1985; Slovic 1987; Asveld and Roeser 2009), and the accident in Fukushima (March 2011) was a major event to test and improve them. A living case of applied philosophy is currently going on in Japan and other nuclearized democracies around the acceptability of the nuclear industry. Will this industry survive the revelations concerning technological flaws, mismanagement, political maneuvers, media manipulation, and the brutal fact of entire regions remaining unsafe for humans for centuries? The empirical data here seem to provide enough proof that the nuclear industry cannot be a way for humans to dwell on the planet. The reasons for continuing the nuclear industry after this accident, and probably after the next one of this type, are difficult to decipher. From the limited viewpoint of ordinary technoethics, a specific point may help to understand the case. It comes from the focus on the extraordinary/ordinary distinction in technology and the primacy of the ordinary in this book. Nuclear industry is emblematic of extraordinary technology and its assessment depends on this status. The assessment of this industry in terms of risk is usually a cost/benefit approach – with the most inclusive meaning for "cost" and "benefit" (political, symbolical, esthetic, long-term and long-distance consequences of all sorts). This field is for experts and both sides hire brilliant experts. The ordinary use of electricity happens on the other end: the user-end of electrical lines. Relatively cheap electric power determines life arrangements and habits in countries deprived of natural power resources (such as Japan) and in countries where the state enforced a high degree of nuclearization (such as France). The ordinary use of air conditioning or electric heating is a subject for ordinary technoethics. On this side of the electric line, the awareness of sustainability and risk issues was extremely low, until recently.

This is generally the case for ordinary technoethics. The question is how to raise the level of awareness and commitment. We must shift from an "extraordinary" question about nuclear safety, where the catastrophe lingers and fosters both ideological technophobia and counter-ideological propaganda, to the treatment of ordinary questions like owning and using air conditioning and heating devices or equivalent appliances in the ordinary technosphere. To make it visual: in Japan, one is easily concerned about the shape of a nuclear plant on the horizon but not about seeing cold-beverage vending machines in the street and everywhere, for instance, or the flashing lights of advertisements. This can lead to effective consciousness and commitment. The argument is not different from the moral argument on climate change (Gardiner 2011): the blockage of sustainability issues in the top-down approach can be forestalled from the bottom, in making the links between global issues and personal microactions explicit and understandable to all. A different perception of the global is thus generated, focusing on the ordinary construction of the global by our ordinary behaviors. A different notion of sustainability is implied as well, focusing on micro-agency.

Sustainable Use

Maintaining, repairing, recycling, and similar activities are the pragmatic outcomes of new societal sensibilities. This style of sustainable use can be mocked as a fashionable new urban style, the famous *bobo* (bourgeois bohemian) lifestyle. A first rejoinder would be to stress that the bobo lifestyle is factually beneficent for the environment and for a sound economy, at least compared to the average upper-middle-class consumer's lifestyle. Further, the bobo is hopefully a source of educational virtue from the elite: if the masses imitate their degrowth and organic style as they have imitated the previous voracious consumer style, this would be a painless reeducation of the masses by social contagion. More seriously, the elaboration of a lifestyle that tries to be systematically conscious of local and global sustainability must be taken as an opportunity and more than a transitory social fashion. This opportunity for change specifically affects the ordinary interface with the technosphere.

The rising popularity and social value of items that are robust and durable, reparable, certified, and labeled for their ecological and social acceptability demonstrate an emerging ordinary technoethical sensibility. Sustainable use modifies the meaning of use in itself. Predatory commodities consumption yields to a kind of well-balanced life accompaniment by artifacts, engaging emotions and particularly attachment bonds. The education and self-education that are needed to support this societal trend are, for instance, the capacity to use (maintaining it, repairing it, upgrading it) a computer or smartphone as long as it works with acceptable performance, instead of replacing it under marketing and fashion pressure. "Acceptable performance" means something definite in this case: it does not mean "state of the art" because not everyone needs a state-of-the-art word processor, car, kitchen blender, etc. One wants homemade apple juice and the device that has been in use for decades in the family carries and transmits different values than the brand new commodity. The environmental consequences of frequently replacing electronic devices by new ones (intensely promoted) could be diminished by increasing one's personal capacity to assess one's real needs and also one's desires.

Therefore, technoethics calls upon the modest but precious moral notions of *satiety* and *frugality*. Satiety means to have enough and its most ordinary case is food intake. A basic education of the self is, or should be, the perception and refinement of satiety, without which eating disorders endanger not only one's physical health but also one's moral condition. Being unable to have enough constitutes a satiety disorder and it is not limited to food. It is a social epidemic concerning every material possession, and immaterial possessions too, starting with money and power.

The capability for satiety starts with the simple awareness that it is an issue. Questions about one's *satiety level* concerning money, social status, possessions, or any other object of desire, are essential to the constitution of the modern self, in all the phases of life and all the details of ordinary life. *Frugality*, then, will not be the capacity to have little, but the capacity to be aware of one's satiety standard and to actually live by it. The ethically questionable person is not the owner of a diesel SUV as such but the owner of it that does not care. There is nothing to say to the oncologist who is a smoker, he/she deserves some sad respect. Ethical sustainability is a non-utilitarian concept rooted in the sustainability of one's self as a consistent and dignified entity. Self-education concerning satiety and frugality was one of the main topics in Seneca's moral teaching. Foucault tellingly interpreted it as aiming for the perfect and complete satiety of the self, "summa tui satietas" (Foucault 2001: 108).

Ethical sustainability transposes the paradox of affluence, as discussed in the 1970s by Ivan Illich or E.F. Schumacher, into ordinary technoethics. These whistle-blowers did not only state the facts of economic and ecological unsustainability, they diagnosed a deeper philosophical unsustainability, a metaphysical crisis in modernity, said Schumacher (1973). Schumacher prescribed Buddhism as the treatment of this moral pathology, not as a doctrinal solution but as consisting in microactions like "planting a tree," his mantra, with a symbolic and not a utilitarian intention.

The economist in Gandhi's team, Kumarappa, published in 1948 the principles of "village economy" under the title *Economy of permanence*. This program for a social and economic revolution is nothing but a vibrant call for the ordinary. Ordinary lifestyle in itself provides a solution for the "less industrialized" countries or communities that do not want to fall prey to global financial institutions. Although this program was not followed up, it remains an inspiration for voluntary simplicity. The notion of *permanence* in Kumarappa is a remarkable early form of ethical sustainability in its application to the ordinary: "What we eat, wherewith we are clothed and what manner of life we lead, all have a bearing, not only on our own lives, but also on the future of mankind" (Kumarappa 1948: 77).

In Buddhism, mindfulness is the essence of awakening (Hershock 2014: 1) and this remains valid in the technosphere. An ethics of sustainability aims at using and possessing material and immaterial things without being used and possessed by them, which is a Buddhist endeavor (Harvey 2000:

197). Deep sustainability then is an economic and political alternative but more essentially it is an ethical alternative.

5.2 Microactions: Post-political and Post-institutional Engagement

Ordinary technoethics envisions the construction of communities from another perspective than the traditional forms of sociality and politics. The intensive agency of billions of "individual users" on the Web builds in real time a transnational and intercultural equivalent of institutions. There lies not only the real economic agency but the real political agency. Focusing on the ethics of ordinary microactions starts social activism over and it means no retiring from the social. This new process starts from the bottom and the deep, from the self. Dismantling the existing institutions (private and public) is in no way the purpose but instead constructing new working communities based on micro-engagements. Weak ties have taken over (Granovetter 1973) and now they are digital. "Micromotives" are in power (Schelling 1978) and now they are digital.

The Concept of Microaction

The concept of microaction is central to ordinary technoethics. Microaction is our interface with the technosphere, factually, and as a consequence it is our main moral agency, ethically. "Everything is miscellaneous" in the digital world (Weinberger 2007), certainly, but this miscellaneous is ethically important. The challenge is to turn it into agency and flourishing. It requires the awareness of a self and a lot of the virtues of the self.

A microaction is ontologically an action. This includes non-action but excludes discourse – except for *speech acts* in their pragmatic dimension (the promise to wash the car, for instance, is an action as a promise but not as car washing). "I vote 'for' Continental Edison's full range of technical interconnections every time I switch on my electric typewriter," wrote Winner a long time ago (Winner 1977: 234). Now when he writes on his computer he is connected to the Internet and he votes for a lot of other entities that do not even have a name for the public (tracking and advertisement "bots" for instance). On a general level, microactions like driving a car for this precise ride or eating this precise steak are ethical engagements that validate the economic and institutional structures supporting these activities. The immediate consequence is that the corresponding non-actions – plus possible alternative actions, walking instead of driving or sharing a car through a digital service, eating something that stands "lower on the food chain" – have exactly the same effect: they only opt out for a different "vote." Winner's expression "vote" here refers to political choice but there is more. "Voting with one's feet" was more than expressing a political option in the times of totalitarian Eastern Europe and similar historical situations. It was an existential dramatic choice. Voting with one's feet when people change the shop they patronize is a terrifying event for marketing departments. Voting with one click on the Web is far easier and probably more powerful now than voting with one's feet has ever been; voting with one's payment card is paramount. Refraining from using one's payment card, or not clicking on the "pay now" button, is the most decisive ordinary microaction – economically, politically, and ethically. It is interesting to remark that it is a non-action and that it is non-confrontational.

The significance of this practice is reinforced in the stalemate situation reigning in sustainability issues and particularly in the climate change issues, as Gardiner (2011) has persuasively argued. Gardiner demonstrates in detail how the climate issue constitutes a "perfect moral storm" that no existing institution can resist. Alongside profit, technology, and law, another force is required for addressing efficiently this kind of issue: ethics. Its neglect has caused a moral and epistemic "corruption" in our capacity for assessment and action, he says. Gardiner's book specifies the ethical turn in the political and social science of climate change. Technoethics proceeds further in the direction of the ordinary. Individual persons have no real power over energy, oil, or nuclear industries, and I suppose they are wise enough to understand that their personal political vote does not impact these entities. But they have a real power over their own mobility, eating, and lifestyle practices, and furthermore they have a real power of influence on their proximal social circle.

Through resistive microactions, the caring capacity of the self then tackles global issues and not only the local sphere. The self's reappropriation of its capacity to care may be the most significant change in moral anthropology. From this perspective, the main ethical issue is not technology and it is not the global either; it is caring. To care for sustainability and to care for the global, one must first be able to care.

Voluntary Simplicity

The most advanced doctrine pragmatically addressing the question of sustainability is perhaps "voluntary simplicity." This movement exists under different versions all over the world, some of them are clearly sectarian and some of them consist in discreet practices accessible in urban ordinary life. I will favor voluntary simplicity on the ordinary level, a simplicity that is non-heroic, non-ideological, and relates to the new forms of wisdom that I plead for.

Duane Elgin's book on voluntary simplicity, an important reference in the field of technoethics, is not far from ideology at times but it still holds as an inspirational and very accessible presentation of this movement (Elgin 1993). Living in an Indian village, Elgin is aligned with Kumarappa's village economy and its Gandhian inspiration (Section 5.1). An even more Gandhian source is offered by Gregg (1936) but with fewer practical aspects and a regrettable tendency to turn ethical resolution into political mobilization of the Leninist type. Merkel (2003) is an engineer who gives the technical specifications and planning for the smallest possible "footprint" on the planet. His approach, "radical simplicity," is evoked in the book through the expression "global living," meaning a globally sustainable personal lifestyle. French philosopher-farmer Rabhi (2007), or Japanese mystic-farmer Fukuoka (1985), remind us of the importance of being able to grow one's own food, a capacity almost totally lost in modernity. Regaining the awareness of this lack, which constitutes a pragmatic and philosophical weakness, is a step toward wisdom, if not toward personal engagement in farming. In all these interpretations, from mysticism to engineering, voluntary simplicity is a typical technoethical *engagement*. Implementing a part of this engagement in ordinary life can be a project for anyone at any time.

Elgin's formulation: "a way of life that is outwardly simple, inwardly rich" (Elgin 1993: subtitle) captures the ethical essence of voluntary simplicity. The end value is not in material degrowth but in existential growth. Nevertheless, the engagement entails having less, less than what affluence offers, with this important distinction: "Poverty is involuntary and debilitating, whereas simplicity is voluntary and enabling" (Elgin 1993: 27). Simple life means, necessarily, "living more voluntarily," says Elgin (1993: Chapter 4). This point characterizes it as a promising form of ordinary ethics and a form of wisdom that is absolutely not regressive or reactionary.

Collaborative Microactions

Microactions respond to a world of disseminated entities and disseminated agency. Human organizations tend to rely on centralization: one group, one leader, and a top-down organization. Democracy and complex forms of social cooperation (including economic markets) have invented *distributed* organization. This other form of order is well represented by client-server architectures in computer networks: a server, which is a master computer, dispatches information and instructions to "clients" who send requests. After centralization and distribution, a third stage is *dissemination*: a multitude of local nodes cooperate according to a largely peer-to-peer exchange protocol. The Internet is the paradigm and paramount of disseminated entities and agency. As the Internet determines our social organization more and more, the basic form of order in our world progressively evolves into a disseminated order. Engagement in microactions is then a disseminated engagement corresponding to this situation. The interesting characteristic of the Internet is the astonishing productivity of its disseminated order – the "intelligence" of Google. It comes from the massively cooperative processes that are running on the network. These are the actions embodying the concept of collaborative microactions in technoethics.

New global engagements result from myriads of micro-engagements. What Benkler (2006) called "social production," the "wealth of networks" that takes over the "wealth of nations" comes from "the role of nonmarket and nonproprietary production, both by individuals alone and by cooperative efforts in a wide range of loosely or tightly woven collaborations" (Benkler 2006: 2). Von Hippel's notion of "user-driven innovation" (Von Hippel 2005) or Bruns' notion of "produsage" (Bruns 2008) refer to this process.

Benkler's book about the Penguins (open source software contributors) versus the Leviathan emphasizes the disseminated aspect of social production in the new economy (Benkler 2011). It gives a pattern for collaborative microactions in technoethics. This pattern addresses the allegation that social coherence may be fragmented in the era of microactions, an allegation that the available data do not support, argues Benkler (2006: Chapter 10).

Tapscott and Williams (2006) dubbed "Wikinomics" this phenomenon and its theory. Wikipedia is the best incarnation of the engagement in collaborative microactions, on the side of contributors, and on the side of users, who validate by their clicks and their micro-improvements the functioning of this type of website. WordReference forums (http://forum.word reference.com) are an even better case of collaborative micro-contributions that create an incomparably useful, precise, and accessible knowledge base for translation issues.

With a set of references that is also mine (Piaget, Kohlberg, Ostrom, and Wikipedia), Benkler focuses the innovation on *engagement*: "new ways of being profitable by *engaging* people, rather than controlling them" (Benkler 2011: 12). But it is not just about being profitable. This engagement falls into the category of care. While building the consistence of online resources, it builds the consistence of the self that accesses and collaborates. Disseminated forms of collaborative care deserve special attention for their ethical importance and not only for their economic weight. There is a "share" button on many Web pages and smartphone apps. In its present state, it is most of the time a fake collaborative tool, designed as a trap for commercial stupidification and often just to track users and get their data. Nevertheless, there is actually a "share" button and not only a "buy" button. It reveals that sharing is a value strong enough to make a function and a button out of it. Starting with this "derelict" existence of the "share" button the question is now to put something real behind it.

There are limits to the logic of networked individualism (Wellman 2001) and it is clear that the spontaneous benevolence of micro-actors cannot be the operating force for social change as a whole. But it gives a new model to treat a problem that Dewey raised so precisely almost one century ago: reinventing the public sphere (Dewey 1927). Modernity needs to reinvent what the "public" is, wrote Dewey. According to his pragmatism, the people and institutions in power must be judged from the consequences of their agency and not from the legitimacy of their power. To regain or to maintain the true spirit of modern democracy we need to reinvent a public sphere. Dewey's claim, one century in advance, strangely sounds like an appeal for the Internet and the global infosphere as a new political agency: "The new age of human relationships has no political agencies worthy of it. The democratic public is still largely inchoate and unorganized" (Dewey 1927: 109); however, he continues, the Great Community will appear when the capacities of modern machinery are in the service of communication, of a massive social communication (Dewey 1927: 184). In a very different tradition of thought, Watsuji's notion of *ningen* calls for a self (a Heideggerian *Dasein* in this case) dialectically merging with the community and whose fundamental activity is questioning (Watsuji 1996: 31). The Japanese philosopher was not in the mood for democracy when writing this, but seeing ordinary human existence as *critical engagement in a community* is essential for improving the present technosphere.

Engagement in microactions reanimates the jammed mechanism of collective and social action. It constitutes the first level on a "local to global" ascending scale. The first social impact is on the proximal social circle, the set of people whom the self can reach by personal contact in real life. The proximal infosphere is key to mutual influence and mutual education. One's clicks and one's diet impact the others, even if they are not directly involved. Taking the stairs rather than the escalator is at first a healthy and ecological sustainable microaction; the self performing this action is also seen by others who can understand and take the same decision; they can be challenged by the exercise, or at least just have the information that stairs can be used by humans (visibly not a common-sense knowledge nowadays). This is education by action. A lot of the changes that occur in our food habits come from the occasional observation of people eating certain things or cooking in certain ways. Lifestyle matters in education and in ethics.

The Limits of Politics and Institutions

Things would be different if only the limits of politics had not been broken by coevolution. The technological structure of our civilization challenges the political structure that has supported human societies since the beginning of large civilizations. The problem is dramatically simple: nation states are the problem whereas nation states bear exclusive political legitimacy for solutions. This basic feature of the problem explains why we are never in a position to take real action: national sovereignty is on the side of the problem and not on the side of solutions. Peter Singer's intransigence is not outrageous here: "A global ethic should not stop at, or give significance to, national boundaries. National sovereignty has no *intrinsic* moral weight" (Singer 2002: 148, and Chapter 6). Transposing globalization and sustainability issues into ordinary technoethical issues responds to the limitation of politics in the current technosphere. National politics with its election cycles and international politics with its summit cycles are both relics of the previous industrial civilization. Nation states are now too big for the local and too small for the global.

Peter Mair's "ruling the void" metaphor expresses the obsolescence of party politics and national politics (Mair 2013). Suggestions for a *stronger* democracy, such as Barber's (1984), and a lot of initiatives, notably in Europe, for citizen participation, may be insufficiently radical. They intend to reform political institutions by a wider and deeper participation. Arnstein offers an interesting ladder of citizen empowerment in public management, from "manipulation" at the bottom up to "citizen control" at the top (Arnstein 1969). There is no doubt that climbing higher on this ladder is advisable for any institution. The real question is about the nature of the regime on the last rung: is it still political domination or does it bring about a disruptive dissemination into micro-agencies?

In philosophy of technology, Feenberg's notion of micropolitics raises the same question. I will argue that it does not take the step out of politics, the lateral step: the transition to post-politics. Feenberg is attentive to the reappropriation of computer technology by the user, from the French Minitel to the Internet of today (Feenberg 1999; Feenberg and Friesen 2011). He considers "technical micropolitics" as a positive asset for the future, particularly in his rather complex theory of a "second instrumentalization" (Feenberg 1999, 2002), which covers reappropriation by ordinary users. But Feenberg is advocating the continuation of politics with the help of micropolitics and not the disruption that is suggested for instance in Foucault's vision of micropolitics. Feenberg is a political philosopher who does not take the post-political step: refocusing on the self and transposing issues in terms of ethics.

The reasons for a more radical move, toward the post-political or, to use a better metaphor, the reasons to take a lateral step, discontinuing politics and trying something different, are not entirely given by the limitation of politics in itself. The post-political that I argue for is more essentially a *post-institutional* regime. The concept of *institution* that I use is maximally inclusive. It includes, beyond political and public institutions, private institutions like corporations and any business entity. For some applications it can even include casual institutions like the family. Institutions are any established structures of power: public and legal, private and market-driven, all of them. They are merging into a single integrated institutional ecosystem.

I adopt the appellation "Tainter's Law" for the principle demonstrated by J.A. Tainter (1988) in accounting for "the collapse of complex societies." Tainter examines the collapse of at least eighteen civilizations throughout history. Independently of the local direct degrading factors, these civilizations essentially collapsed under the weight of their own complexification. The basic process is exactly my argument in this chapter: what is taken as the solution for the trouble *is* the real trouble. Societies that are already too complex and costly in resources react to a threat (ecological, external or internal political threat, or any catastrophe) by adding new layers of complexity and costs in resources, until they finally asphyxiate. What continuously deteriorates is the marginal productivity of increasing complexity (Tainter 1988: 89–90), particularly the declining marginal returns of institutional complexification. Tainter remarked, decades ago, that we experience declining marginal returns in "agriculture, minerals and energy production, research and development, investment in health, education, government, military and industrial management, productivity of GNP for producing new growth, and some elements of improved technical design" (Tainter 1988: 211).

Gardiner (2011), as we have seen (Section 5.1), confirms and worsens this diagnostic as it applies it to the current issue of climate change. Jamieson's recent analysis of our common failure in respect to the climate issue (Jamieson 2014) is cruel for every stakeholder in this matter and it can be read as an ominous invitation to generalize: we have been so bad on this issue that the worse can be expected for the next ones. Fischer (1990) expands an equivalent diagnostic to "technocratical ideology" as a whole (Fischer 1990: 8). He suggests a "participatory expertise" (Fischer 1990: Chapter 14) that would be a collaborative but still institutional solution, in the traditional acceptation of politics and institutions, with a touch of Frankfurt School as in Feenberg.

Gardiner is still more disruptive, apparently: "If the attempt to delegate effectively has failed, then the responsibility falls back on the citizens again – either to solve the problems themselves, or else, if this is not possible, to create new institutions to do the job" (Gardiner 2011: 433). But in reality the "institutional spell" is still active: when institutions cannot do something create new institutions to do it! We are bound to crash again on Tainter's Law. Gardiner is nearer to making a lateral step when he states: "The whole idea that citizens might be politically responsible for the behavior of their institutions is in some respects a radical and demanding one" (Gardiner 2011: 434).

The step is taken by the new social actors (Section 5.1) that have invented the brilliant slogan "Not in our name!". The reappropriation of political space that was formerly delegated to institutions is a decisive step toward the post-political.

Bureaucracy is nevertheless the distinctive modern institution. It shapes our societies and our lives as if bureaucratic rule were the natural outcome and the natural governance of a technological society. Historical and sociological surveys tend to confirm this bad new (Graeber 2015). But empirical facts of ordinary life and abstract principles such as Tainter's Law converge to raise a growing suspicion about the pertinence of this mode of organization, to put it mildly. In every case where bureaucracy is perceived as a non-acceptable form of domination, and they are numerous in modern society, an alternative way of social organization is needed. More and more often it is found in the infosphere. In academic life or in the job market, in more and more varied ordinary activities, an informal digital organization bypasses the established bureaucratic way.

I propose the term *institutional glue* to name this phenomenon: when an institution is involved in a project, the work flow and the timeline tend to slow down and get meaninglessly complex, with consequences that range from moderate inertia to total paralysis. The magic effect of post-institutional actions, typically those on the Internet, can often be explained by the simple absence of institutional glue: just do not glue it and it works.

Post-institutional and Post-political Action

Elinor Ostrom draws a model for organizing human communities from the existing bottom-up management of common natural resources. This model is post-political in the sense that I mean it because it is grounded on local and inter-individual relationships that owe nothing to partisan politics and because its reason for being is not to realize a predetermined form of society or human life (ideology) but only to wisely manage a local common resource (pragmatic ethics). It can be called micro-institution or bottom-up institution; technology now supplies the perfect tools for this form of community.

The new form of social action described by Manuel Castells can also be called infra-, micro- or bottom-up-politics; the lesson to be learned is that no institutionalization process is at the center of the new social demand: these are movements without definite demand, movements where "the process is the message" (Castells 2012: 185).

Post-institutional and post-political actions are such because they belong to existential politics. They engage individuals and communities on the path of the third form of power (Table 1.3), the power over oneself. For this reason, their minimal claim is autonomy. The second form of power, the domination that fuels institutional and political entities, and also the first form of power, technology, that frames the plans of "technical" solutions, are both discarded. Creating "togetherness" (Castells 2012: 225) is already an end in itself for these social movements. Digital togetherness has an amazing capacity to replace the political and institutional forms of being together and acting together. "The horizontality of networks supports cooperation and solidarity while undermining the need for formal leadership," says Castells (2012: 225). Technology supports utopian aspirations, which come from the 1970s, but now flourish in a new environment: the global infosphere, progressively replacing institutional structures. A lot of the applicative virtues in wisdom ethics (Table 1.4) are present in these "highly self-reflective movements" (Castells 2012: 225), both at the level of the collective and at the level of the individual: awareness, autonomy, courage, and in the most advanced movements there is also humility (no leader), harmony (merging the movement into the social texture and then inducing long term effects), and benevolence (in the internal sharing practices of the community).

The capacity of these new communities to merge with existing institutions (Ostrom 1990) is important for the constructive perspective in technoethics. The new forms of agency afforded in the technosphere, particularly in the global infosphere, are not only pragmatically efficient (people really find jobs and sell second-hand books and bicycles, for instance) but they create *norms*. The emergent normative capacity of disseminated microactions characterizes the post-institutional alternative.

The Internet is once again both paradigm and working tool. Wiktionary and Google are the best testimony of the real state of a given language at a given moment in time. I bet that I am not the only scholar who gradually shifts from academic and prestigious dictionaries to collaborative websites just because the words that are really used and the precise meaning they have (so quickly) acquired belong to another timeline, the connected timeline. Their normative capacity does not come from their being *better institutions*, which better "dominate" the subject, but instead it seems to come from their *not being institutions*. They do not follow a logic of domination and mastery but rather a logic of open participation and real-time collaborative improvement. This process, simply, is starting to be the new normative agency. An ordinary language fact goes global through the simple contribution of an anonymous self to Wiktionary. These microactions create a global expertise that surpasses any expert. A fascinating demonstration can be tested on WordReference forums (http://forum.wordreference.com), where "ordinary" scholars and students share about languages. On these forums, some contributors are not experts and some are; online literacy and practice allow detecting most of the time the real expert's hyper-precise answers and suggestions. I know of no equivalent to these forums in a book or human person.

Shirky (2008) offers many illustrations of this "power of organizing without organizations" as he puts it. The apparent paradox dissolves with the idea of post-institutional structures. Organization is possible without dominating structures and this capacity emerges from a new agency: from the empowered masses of individuals. It is the typical emergence in the digital age: the intelligence of the global infosphere. Surowiecki (2004) coined it the "wisdom of crowds" and there followed a "crowd-" hype in business literature. In this matter, the focus should not be on the crowd as such, but on the connected individuals that compose the new crowds: a "crowd" is the connection of *ordinary* selves, this is not a detail. Surowiecki's message stresses the importance of creating the conditions for the emergence of the best out of massively connected ordinary selves.

Micropolitics after the ethical turn can be found in a vast variety of initiatives that have happened since the 1990s everywhere in the world and in every sector of human activity. Ekins (1992) or Epstein (1996) offer a picture of the micropolitics trend before the full reign of the digital. Beerbohm (2012) explains the role of ethical microactions for renewing democracy. In all the examples they give of grassroots movements, the logic of reappropriation and generativity is central. There appears to be an obvious continuity with the explosion of the Web in the last years of the twentieth century. Some of Ekins' case studies mention sensitive issues, for example the Women Living Under Muslim Laws association (www.wluml.org), where the micropolitical question is totally ethical and totally "ordinary" at the same time. In similar cases, the confrontation of collaborative networks of courageous individuals with institutionalized domination gives sense to post-political issues. This confrontation is possible only through the disseminated network of the infosphere. Ekins' conclusion concerning global issues like the environmental crisis is not surprisingly akin to Gardiner's idea of un-delegation: "Once again it has been ordinary people working through largely voluntary organizations who have acted decisively for human well-being, while the established power structures were either blind to the perils or actively promoting them" (Ekins 1992: 165). Since the time of "systemic" psychology, the importance of contexts and environments is recognized in human and social studies. Recent research in applied ethics insist more and more intensely on the impact of moral environments on individual behavior, particularly in virtue ethics (Appiah 2008) and in the ethics of technology (Verbeek 2011: Chapter 6). Focusing on ordinary actions leads to appreciating the technosphere and infosphere as active moral environments. The circumstances of ordinary action are today largely linked to the technosphere, directly or indirectly. The idea of "persuasive technology," which would moralize human (technological) environments instead of directly moralizing people, has been suggested among others by Hans Achterhuis (Verbeek 2011: 95). This leads directly to technocracy and paternalism if the responsibility for designing the technosphere is left to designers alone. For this reason, the user's empowerment to shape and co-design the technosphere and infosphere is imperative in technoethics. Persuasive technology means top-down manipulation as long as the paradigm remains institutional domination. When the environment-shaping process is part of a post-institutional collaborative agency, it carries different values.

The situation may seem similar to the "missing masses of morality" question (Sections 1.1 and 3.3) but it is not. The post-institutional micro-agency that I try to isolate here is more than a displacement of agency as in the "missing masses" paradigm. It is the emergence of *a context-creating agency*: it morally shapes the environment that morally shapes our behavior. Examples in ordinary ethics are numerous, some of them material (the alcoholic beverages that I decide to buy and stock at home, or not, create a moral context) and some of them are digital (my benevolence in answering an awkward question on a forum creates a moral context). The technosphere and the infosphere as a whole are moral contexts, globally. In them, microactions collaborate to shape emerging moral contexts.

"Nudge theory" offers principles and examples of this course of action, most of them in the material world, which is in fact the technosphere. The example of food display in a cafeteria (Thaler and Sunstein 2008: 3) shows how a wise "choice architect" designs a local portion of the technosphere. "Nudge" is defined in the following manner: "[A]ny aspect of the choice architecture that alters people's behavior in a predictable way without forbidding any action or significantly changing their economic incentives" (Thaler and Sunstein 2008: 6). Even when used by institutions it represents a post-institutional way of action, through the creation of contexts. The technosphere and the infosphere are not only contexts and mediation, they are nudging environments. Personal microactions are often nudged actions and awareness can turn them into nudging actions.

Resistance, Resilience, Self-reliance

The conclusion of a daring real-world experiment in digital politics can be impressive: King et al. (2013) have tested what kind of message is censored on the Internet by the Chinese government and they have found that what is considered dangerous is not ideological or even personal "critique," but calls for action, calls for precise action in the real world. The emergence of a real-world agency coming from online communication is what is feared by oppressive governments: "Our results offer unambiguous support for, and clarification of, the emerging view that criticism of the state, its leaders, and their policies are routinely published whereas posts with collective action potential are much more likely to be censored" (King et al. 2013: Abstract). In China, collective action is feared; "critique" not so much. Institutions are aware of the difference between the realm of discourse and the realm of action. Online agency in this kind of situation is initiated by self-reliant individuals. The infosphere allows a typical propagation and aggregation of agencies and eventually the real world is impacted when the protestation comes to real-world action.

In the USA, Aaron Swartz's resistance to digital governmentality was intended as civil disobedience, influencing the institutions through non-institutional action: hacking and promoting its cause online (Knappenberger 2014). The authorities perfectly understood the danger of this resisting self and prosecuted (persecuted) him appropriately (leading to his suicide). Online personal protestation, initiated by the self-reliance of a bold individual, turned into public action in the real world and eventually had an influence on the institutions: to stop the SOPA bill in 2012 (Stop Online Piracy Act).

Even when there is a charismatic leader, new social movements rely on anonymous agents, online and in the streets. But is it "ordinary" ethics? Malcolm Gladwell's paper against the efficacy of "small changes" in 2010 (Gladwell 2010) maintains that heroic individual engagement is necessary for real social change, as in the Civil Rights movement in the South of the USA in the early 1960s. But that was the twentieth century, a century of leaders and followers, for the best and for the worst. The twenty-first century can be a century of collaborative anonymous with a different form of power. In the past, the efficiency of the Civil Rights movement was communicational, through TV and through the printed press, as was Gandhi's (since his first actions in South Africa, relayed by local and British press). These movements would have been more efficiently helped by present day digital media. The press coverage during the 2014 protest in Hong Kong (Occupy Central), where the mass media were only relaying the online vibrant infosphere, testifies for this new order of political mediatization. We have seen with Castells how the new social movements are linked with digital media; Gladwell's article was typically written shortly before these movements. Small changes matter in the present technosphere and infosphere; they can initiate deep change.

A new social resilience and resistance emerges. The post-political in it is a provocative engagement and not a disengagement. The depoliticization of issues, the lateral step, is a new form of engagement that disrupts the established consensus in public affairs. This engagement initiates resistance on a deeper level than oppositional politics. The most important disruption is not in the confrontation of discourses but in the change of level itself: the fact that resistance is not a discourse but a course of action. People do not demonstrate in the streets against the economy of show-business and its cultural consequences: they simply download from the Web and exchange among ordinary people the music they like, independently of copyright monopolies and of the cultural choices of mainstream media.

Social, political, and economic resilience lies ultimately in the self. The resistive self is the pivot point for ordinary technoethics. Foucault dared say in his last conferences: "There is no point, first and ultimate, for the resistance to political power, other than the relationship of the self to the self" (Foucault 2001: 241).

5.3 Inclusive Ordinary Technology

Theory of Inclusive Technology

The shift from valuing "exclusive" to valuing "inclusive" encapsulates the emerging ethics of the ordinary that I am trying to articulate. "Exclusive" denotes something of value, a luxury item or service, something that people are supposed to crave for, then to indulge in as soon as they can afford it, or to envy if they cannot. The idea of *exclusion* as such remains secondary and apparently benign. It is only as a consequence of the value (the cost) of the thing that some persons are excluded and resent it. In fact this exclusion is an integral part of the value of the thing in a consumer competitive society. For some people, the pleasure in owning that car or smartphone is that most of the others cannot afford it. As soon as they can, the object loses its exclusive value. It has happened to cars or mobile phones.

Vanity is a prime moral flaw to be taken into account if we want to decipher ordinary behavior and reorient it toward wisdom virtues. The virtues of satiety and frugality and the whole set of wisdom applicative practices (Table 1.4) plead against vanity and exclusiveness. Even in this case modern wisdom will not adopt a confrontational stance against vanity and exclusiveness. The positive attitude consists in promoting inclusiveness, which leads the debate on a lateral and alternative course of ideas and actions.

Inclusive means "not excluding" in the common use of "inclusive technology," which is particularly appropriate for ordinary technoethics.

The MS-DOS operating system was relatively exclusive (selective), recent Linux, Windows or Mac operating systems are inclusive. The smartphone has been such a success because it is amazingly inclusive, at least for digital natives and for reasonably motivated digital newbies.

Universal Design (UD) is the name of a large movement for inclusive design and technology, particularly active in Northern Europe and in Japan. It is also known as "design for all" or "barrier free." National legislations, all over the world, progressively incorporate UD concerns and UD norms. The current physical technosphere excludes not only people with disabilities in the strict sense of the term but also "regular" people from the moment they carry heavy baggage, are pregnant, or cannot read the local language, for instance. Most of the "regular" persons, anyway, will age and lose some of their capacities – physical, perceptual, cognitive or other. A striking slogan of UD reminds us that every human is not a young (rich) Anglophone male in perfectly good health – although the world seems to be made for this minority. This is probably because it is shaped by this minority.

A different logic is at work, as the orientation signs in international airports demonstrate for instance. They use pictograms, images, and not language for guiding travelers in a complex environment. A lot of electronic devices try to use the same communication logic in order to provide a fluid and comfortable "user experience." This is a growing trend in the major fields of technology. It belongs to the simplification law above (Section 2.3), which offers strong support for UD (Herwig 2008: 168–169).

Making the technological future "easy for all" presupposes a deliberate effort to plan and to design for the ordinary. Air travel and computer usage being no longer extraordinary privileges, they need to be facilitated for everyone. The ordinary technosphere demands universal access: in this sense the efforts of those that make things simple and maintain them so should be praised almost as much as the genius of first inventors. Being inclusive is a core value of modern technology.

A striking parallel is possible between the claim for *justice* in a context of material abundance and the claim for *access* in a context of informational abundance. Immaterial affluence in the infosphere (e-affluence) makes access a fundamental right for human persons. A very large notion of access must be defined: access to places, material and immaterial, access to data but also access to culture, access to persons, and finally free communication. In the decade between his book about "the age of access" in 2000 and his book about "the empathic civilization" in 2009, Jeremy Rifkin has punctuated in his own way the expansion of access in modernity. The present state of affairs in the immaterial (digital) world is characterized by the value of access as such. The debate about "the neutrality of the Net" (equal access for every user to every server and content) is not only resistance to the commodification of the commons: it touches the core issue of access and its value as a principle. Electronic connection must be inclusive. The emerging values depend on this openness.

This tension is often active under a resistive guise: the non-acceptability of non-access. It is well known in digital security circles that if you restrict the access to a part of your system you are immediately met with a huge amount of resistive energy: there will be people prepared to work all night long just to access what you want to protect, for no other reason than the pure right of access, or better the non-acceptability of non-access – this complex resistive value may be called simply "fun" by hackers. When one receives a pdf file with a restriction in the rights you have to print it or copy it, one may feel an impulsion to look up some shareware in a forum for unlocking that restriction. Just for "fun," in a sense, but something more important lies behind attachment to free access and use.

Access is an excellent candidate for being a transversal value, both in the real world and in the infosphere. It is also an efficient trigger to mobilize social energy, a tragically rare and precious resource these days. The feeling of legitimacy for personal or public action in the defense of access, which is a feeling of the unacceptable, points toward a deep evolution of our common values. This new set of values could be called "Scandinavian," as it reigns in that region of Europe more than in any other place. A kind of "Scandinavian spirit" might now be taking over the "Californian spirit" at the origins of neo-technological culture. A new utopia is taking shape in continuity with the Californian spirit that inspired a first wave. This second wave goes deeper into the ideas of sharing and universal flourishing as it insists on inclusiveness in technology.

Four emergent e-values characterize the most promising forms of inclusive technology: facilitation, transparency, collaborative processes, and end-user design. The classical seven principles of Universal Design (http://universaldesign.ie/exploreampdiscover/the7principles) give a simple pattern for inclusive technology and e-values. Equitable Use (1) and Flexibility in Use (2) are easier to design in the infosphere than in the material world; Simple and Intuitive (3) use, with transparent and always available Perceptible Information (4), are central for e-designers; Tolerance for Error (5) is possible and is standard in electronics more than anywhere; Low Physical Effort (6) and appropriate Size and Space for Approach and Use (7) are made considerably easier by the dematerialization of the processes, on the one hand, and they are addressed as priorities in the material interface of e-devices, on the other hand.

The "Scandinavian" spirit, through inclusive technology, opens up a world explicitly designed for a broader range of human persons. The technosphere and infosphere are more and more constructed by ordinary users, a broader range of people than ever. To push the idea to the limit, let us imagine an environment mostly designed by old (poor) sick ladies from Asia or Africa. It would probably terminate the automobile civilization, for one thing. It would insist on care more than profit, I guess. This design method would shake more than the car cult and the profit cult, it would hit the ideological barriers that paralyze social evolution.

Inclusive technology tends to implement an ethically *integrated* techno-sphere and infosphere, basically different from the confrontational universe brought about by usual politics and business practices, and intrinsically different from the systematic confrontation created by national, local, ideological, or religious enclosures in geographical and cultural space.

The Question of Design

A strong tradition in philosophy of technology has stressed the importance of design as the phase in which values can be infused in technological arti-facts. This concern bears more and more on the inclusive aspects of *ordinary* artifacts through the priority given to "user-end" design or "par-ticipative design" (Friedman 1996; Mitcham 1997; Oudshoorn and Pinch 2003; von Hippel 2005). The "fine grain" of usability is important and most of the time it implies ethical considerations.

Verbeek's (2011) provocative formulation, "designing the morality of things," proves that design is central but in a new configuration: *value design* in the contemporary technosphere does not mean the omnipotent domination of a prior theoretical expertise. It claims to represent the user's best interest as it is expressed directly by the user. An ironic "tyranny of users" can be felt in a lot of design departments. In the complex and moving patterns of constant exchange between end-users and designers, as it exists in modern industry, the balance is now largely in favor of a domi-nation by users and customers. *Including* their expectations and reactions in the product has initiated an inclusive design methodology – where the motivation remains pure profit and marketing advantage. This effort nevertheless coincides with the ideals of inclusive technology. Gems of usa-bility (some smartphones or websites) are due to a realistic economic motivation on the side of corporations. It may be morally sad but the important fact remains the high level of inclusive technology now in use in the technosphere.

A minimal understanding of ethical issues belongs to the basic skills for contemporary engineers, especially according to European perceptions of the social role of the engineer (Goujon and Hériard Dubreuil 2001). The loyalty of the engineer is less exclusively to a firm, as compared with the USA, and more to society at large or better to the ideals of society, its abstract norms. Recent studies about "Values in engineering design" (Van de Poel in Meijers 2009) indicate that trade-offs in design and value con-flicts are more and more frequent in engineering. "Value conflict is in fact at the heart of the design process" (Van de Poel in Meijers 2009: 1004). "Value Sensitive Design" makes sense in more and more engineering con-texts (Tavani 2004: 125–127). The moralization of the technosphere occurs in a context of pervasive designing where responsibility and initi-ative are disseminated among multiple agents and multiple actions. The danger of paternalism is warded off by the disseminated character of designing agency. The new agency that I try to capture in this book, a solid self engaged in self-care, can meet morally designed technology and eschew the paternalistic model because this self is itself a robust designer of its own life. Techniques of the self, meaning wisdom practices, prepare the adequate subject for a technosphere where objects are morally designed. This subject can accompany or resist these objects.

In the end, the key feature of innovation in the present technophere is again *generativity* (Section 4.3) in the sense of Zittrain (2009). Inclusive technology leads to a pervasive generativity: in every computer or smartphone the human self evolves a specific environment that connects with the other personal and common environments of the infosphere and the technosphere in order to create the conditions for flourishing. This bottom-up generativity bestows upon inclusive technologies the value of *resistance*, in the sense of Wu (2011): on the digital battlefront, for the first time in history, the tragedy of concentration and disappropriation of the media in favor of a dominating elite is doomed. In front of commercial and bureaucratic exclusion there emerges everywhere collaborative inclusion. In front of every Microsoft, there tends to be a LibreOffice.

The Soft Power of Inclusion

Inclusive technology is destined to be active in every cell of the social body. It is *ordinary* technology because it is disseminated in each and every micro-sphere of human and social existence. At first sight, large international movements and programs like Universal Design look like an institutional constraint imposed on social actors and individuals. Under this interpretation, UD will easily remain a formal and bureaucratic inclusion policy, with no deep effect on reality. Real implementation begins when, in application of official programs or emerging from the implicit values, inclusion becomes natural. *Soft inclusion* is analogous to and in fact uses the channels of the *weak ties* in the new social order, theorized by Granovetter (1973). People find new jobs as they find second-hand items: through the informal networks of a disseminated economy and a disseminated sociality, not through public institutions or private institutions (corporations and market). A pervasive support system is available for everyone in the infosphere, easily accessible – just knowing a URL or a site name, just being able to write an astute Google request. For technoethics, this daily support and reassurance is of prime importance because it typically takes place in the ordinary, as soft inclusive patterns in everyday life. Underneath "hard" established social structures, the soft power of inclusive and collaborative networks is continuously reweaving and repairing the social fabric.

The disappropriation of the skills necessary to build and maintain a house was denounced by Ivan Illich (Section 2.4). A partial reappropriation of house building and maintenance (electricity, plumbing, heating devices, etc.) is available on the Web to remedy this impoverishment, usually under the very easily assimilable form of video films with an excellent informal pedagogy. Some of these tutorials are due to professional networks that intend to sell the tools and materials involved and some are due to enthusiast amateurs, but the difference is of no importance – or better: the competition for audience may explain the excellent pedagogy of the videos.

With this support, an individual, even in relative geographical or social isolation, can embark on an ecologically sound home and lifestyle. This person, if sufficiently motivated, will find online, and will acquire through online orders, the materials and devices to carry out her project. In a real and symbolic sense altogether, inclusive dwelling on the planet is made possible. Globally, it engages a new vision of technology: it is high-tech (the Web and the devices to access it) but driven by a personal value project and not delegated to a cast of professionals. Inclusive technology here does not mean the replacement of (hard) professional expertise by mediocre amateurism but a completely different landscape: expertise disseminated in a non-dominating (soft) mode.

One of the challenges in the current phase of this trend is the extension of its virtuous logic from a purely personal level ("my own home and lifestyle are ecologically sound, may the Deity help the others!") to a collective and social level. How will urban planning, for instance, shift from its current regime (domination by institutions, politicians, and corporations, except where the Scandinavian regime holds) to the practices of reappropriation by a collective agency? We must find how possible virtuous micro-actions and micro-projects can diffuse and expand this form of soft empowerment to the collective and the social.

A massively inclusive digital socializing already reigns in the life of almost every young person in the present infosphere. Social networking sites, but more than anything the diverse and personal social networking ordinary activities, have taken over the essential function of integrating young people into society. It is the case for social inclusion into their own age cluster and progressively for their inclusion into society at large. For the first time in history, a cultural immaterial sphere, the infosphere, is the public place where the decisive phase of social inclusion largely takes place. Danah Boyd has given splendid accounts, at the same time critical and appreciative, of this phenomenon (Boyd 2008, 2014). The key issue, once the importance of this fact is acknowledged, is to create the conditions that foster an authentic practice of digital social inclusion.

The ethnographic approach of American teenagers' digital sociality in Boyd (2008) justifies the notion of *engagement*. The late 1990s and early 2000s were a time of transition between the era of MySpace and the era of Facebook. More importantly, at that moment emerged a new form of social engagement online. It almost instantly proved to be a *focal engagement*, in the sense of Borgmann. As far as the content is concerned, the shift was from a music- and leisure-centered network (MySpace) to a person-centered network (Facebook). In fact, it was from a recreational engagement to a deep intimate engagement. Online activities no longer create a particular community of committed persons but it extends to the overall community of humans. In Facebook and similar immersive and inclusive social networks, the de-specialization of the content corresponds to the existential significance of social networking. Some of the first networks were oriented toward dating, actively meeting people within a specific context (seduction), which is typically a one-dimensional existential engagement. With Facebook and similar "generalist" networks, the scope is existence in its entirety, including seduction but also job seeking, pure fun, cultural projects and so on. What is engaged in this focal activity is the person in its entirety.

The technophobic misinterpretation of digital social networking as "excluding" people from "real life" is particularly prejudicial to the understanding of modernity. This blockage is even fatal for an appreciative consideration of the essential capacity and mission of ordinary digital activities: soft inclusion. The reasons for this blockage are quite obvious: it is about power once again. What is now delivered by online activities, the social inclusion of young people, is a place of contest for power. It was formerly under the domination of parents, schools, churches, and nation states. It evades these domination powers and not unexpectedly meets fierce resistance from all of them. Ordinary technology as it already exists reveals vested domination frameworks. Evolving them into harmonious collaboration and non-dominating interdependence is a task for technoethics.

According to Boyd (2008), three dynamics are at work in this new online sociality: invisible audiences, collapsed contexts, and the blurring of public and private. Teenage sociality, from intimate (family level) to distant (institutional level), is mediated by digital networking which has become an interface. This interface imposes new proximity and distance relationships. There is something figuratively "spatial" in the idea of inclusion: it is apparent in the spatial metaphors of online "sites" and "presence," the complex topography of "links," the revealing persistence of "traces," and all the "moves" in online space starting with "surfing." Through these activities, teenagers discover and experience the world, including the human and social world (Boyd 2008, 2014).

The new location of inclusive processes, the infosphere, has heavy consequences. As technoethics has noted (Ess 2001, 2009), cultural globalization deeply affects one of the most intimate constitutive structures of human mind and society: it "flattens" the infinitely rich and diverse existential repertoire of mankind. Here, Hershock (1999) identifies a form of colonization of the minds, which is a real concern.

The famous "fear of missing out," the FOMO syndrome (always checking one's messages, profiles, accounts) is not exactly a pathology or an exaggeration. It touches a key preoccupation, social integration, which is now happening through the inclusive channels of the infosphere. It does not mean a stupid addiction and intoxication by a gadget; it embodies a Kierkegaardian existential anguish and need for reassurance. Identically, there is nothing despicable in itself in the idea of "writing oneself into being" (Boyd 2008: 121): wasn't it a privilege formerly reserved for some artists? It is now available to any connected teenager, who is probably not a Marcel Proust I must concede.

Boyd's recent book (2014) evokes the last steps in online sociality. It emphasizes the typical ambiguity concerning *authenticity* in digital mediated experience. On the one hand, ordinary social digital networking happens in the private and intimate part of life (The Celebritization of Everyday Life, Boyd 2014: 147). On the other hand, this ordinary intimate becomes public and published (Boyd 2014: 201).

The strange couple status indication on Facebook, "it's complicated," is in the title of Boyd's book. It is emblematic for social networking as a whole: it is ambiguous and often embarrassing. "Reality is nuanced and messy, full of pros and cons. Living in a networked world is complicated" (Boyd 2014: 16). I suggest that this ambiguity has something to do with the deep level of *ordinary* that is involved in digital social networking. Nothing is closer to the essence of the person, its constitutive choices and frailty, than details about the miscellaneous events of the day, the detailed recording of emotions, encounters, micro-experiences of the day, as they are confided, in the middle of the night, to the distant and invisible audience online.

What happens, then, with great moral importance, in digital socialization does not concern technology or anything new brought on by technology. It concerns the social inclusion of adolescents, a perilous subject for parents and educators. Since it is transposed in the infosphere, and because teenagers' psychological intimacy is frequently an embarrassing subject, a temptation looms to transform the initial preoccupation into a preoccupation about technology and its novelties. It is not easy to accept one's anxiety about who one's young daughter is going to meet and make friends with in real life, but since it happens online, a good opportunity to feel more comfortable or consensual with this anxiety is to transform it into an anxiety about who she is going to meet and make friends with *online* – a typical case of technological scapegoat. Educators should fully acknowledge the infosphere as a medium for sharing pain and finding real support (Boyd 2014: 123). The healing and assuaging power of digital media remains largely unaddressed. There is yet a potential here to improve our civilization on one essential point, concerning intimacy and social inclusion.

Forced Inclusion: Hacking and Piracy

A last and provocative application of inclusive technology may be found in the practices of hacking or piracy. I broadly define them as any intrusive, unauthorized access or use of an element of the infosphere through technological means. I suggest an interpretation of hacking in both its forms, innocent and malevolent, as a mode of inclusion. This does not prevent condemning malevolent hacking because it is malevolent, for its intention then, and not because it is hacking, for its technological method. Hacking can be an instance of post-political action and a paradigmatic one as it is an instance of disseminated inclusion through the infosphere, and a paradigmatic one again. This follows from the primacy of the ordinary, which qualifies all of us as so many humble hackers, acting on a post-institutional scene, networking in an informal collaboration pattern that is post-political and post-economic. The general idea of "hacking the technosphere," in a constructive and non-confrontational way, can be a sort of maxim for technoethics. The idea of a wise "hacking" of the world makes sense: philosophers have always *hacked* the cultures that were magnanimous enough to tolerate their presence.

In its historical approach to hacking, Steven Levy's early book (1984) presents it as being simply a set of behavioral choices. These choices in fact are values, ethical values: for independence, self-reliance, and beauty. Levy concludes with a question: why did the hacker's image evolve into that of delinquent? The answer may be found in the natural coevolutionary "biodiversity" of a generation that produced both Steve Jobs and Richard Stallman, the champion of business and the champion of anti-business. Coevolution produced millions of similar non-identical twins like these two, millions of anonymous contributors to Linux or LibreOffice on the one hand and thousands of malevolent programmers on the other hand, exactly as inventive and uncontrollable as the rest.

Himanen (2001) is the authority on hacker's ethics. He gives substance to the idea that the prime value is *fun* for a lot of hackers. Actually, even in its delirious and deviant form of "lulz" (provocative chaos for no reason at all) the fun in hacking's motivation explains the *inclusive enthusiasm* that leads to intrusive practices or exuberant self-promotion.

Recently, in declarations linked to Wikileaks or E. Snowden, the news was filled with hackers' fierce militant and political "mission statements" (resistance to universal surveillance and the abuses of established powers). This discourse may be a strategy to gain some acceptability for hacking practices that are officially considered threats to "national security." Are Assange and consorts legitimate whistle-blowers for democracy and freedom? The politically simplistic background of their statements, which stand on the edge of pure conspiracy theory most of the time, does not plead in their favor. But let us compare the *extraordinary* and self-promotional activities of these iconic hackers with the humble downloading, unauthorized uses and customizations, illegal sharing practices, and so on, of millions of anonymous *ordinary* Internet users. The *anonymous*, without the capital letter, are more important than the revered Anonymous – and authentic Anonymous militants, if I am not mistaken, agree with me on this point. It is a decisive point in the constitution of a sapiential technoethics: humble disseminated microactions matter the most. It applies everywhere, as it does here in the interpretation of hacking as a deviant mode of inclusion.

References

Appiah, Kwame Anthony. 2008. *Experiments in ethics.* Cambridge, Mass.: Harvard University Press.

Arnstein, Sherry R. 1969. A ladder of citizen participation. *Journal of the American Institute of Planners* 35(4): 216–224. www.apho.org.uk/resource/item.aspx?RID=82367.

Asveld, Lotte, and Sabine Roeser (eds). 2009. *The ethics of technological risk.* London: Earthscan.

Barber, Benjamin. 1984. *Strong democracy: Participatory politics for a new age.* Berkeley: University of California Press.

Beerbohm, Eric. 2012. *In our name: The ethics of democracy.* Princeton, N.J.: Princeton University Press.

Benkler, Yochai. 2006. *The wealth of networks: How social production transforms markets and freedom.* New Haven, Conn.: Yale University Press. www.benkler.org/Benkler_Wealth_Of_Networks.pdf.

Benkler, Yochai. 2011. *The Penguin and the Leviathan: How cooperation triumphs over self-interest.* New York: Random House.

Birnbacher, Dieter. 1988. *Verantwortung für zukünftige Generationen.* Stuttgart: Reclam.

Boyd, Danah Michele. 2008. *Taken out of context: American teen sociality in networked publics.* PhD Thesis. Berkeley: University of California. www.danah.org/papers/TakenOutOfContext.pdf.

Boyd, Danah Michele. 2014. *It's complicated: The social lives of networked teens.* New Haven, Conn.: Yale University Press. www.danah.org/books/ItsComplicated.pdf.

Brundtland, Gro Harlem (ed.). 1987. *Our common future.* www.un-documents.net/wced-ocf.htm.

Bruns, Axel. 2008. *Blogs, Wikipedia, Second Life, and beyond: From production to produsage.* New York: Peter Lang.

Castells, Manuel. 2012. *Networks of outrage and hope: Social movements in the Internet age.* Cambridge: Polity.

Cooper, John M. 2012. *Pursuits of wisdom: Six ways of life in ancient philosophy from Socrates to Plotinus.* Princeton, N.J.: Princeton University Press.

Dewey, John. 1927. *The public and its problems.* New York: H. Holt.

Douglas, Mary. 1985. *Risk acceptability according to the social sciences.* London; New York: Routledge. Repr. in *Collected works,* vol. 11, Routledge, 2003.

Ekins, Paul. 1992. *A new world order: Grassroots movements for global change.* London; New York: Routledge.

Elgin, Duane. 1993. *Voluntary simplicity: Toward a way of life that is outwardly simple, inwardly rich.* New York: William Morrow, 1981. Revised edition.

Epstein, Steven. 1996. *Impure science: AIDS, activism, and the politics of knowledge.* Berkeley, Calif.: University Press.

Ess, Charles (ed.). 2001. *Culture, technology, communication: Towards an intercultural global village.* Albany, N.Y.: State University of New York Press.

Ess, Charles. 2009. *Digital media ethics.* Cambridge: Polity Press.

Feenberg, Andrew. 1999. *Questioning technology.* London: Routledge.

Feenberg, Andrew. 2002. *Transforming technology: A critical theory revisited.* Oxford: Oxford University Press, 1991. Revised edition.

Feenberg, Andrew, and Norm Friesen. 2011. *(Re)inventing the Internet: Critical case studies*. Rotterdam: Sense Publishers.

Fischer, Frank. 1990. *Technocracy and the politics of expertise*. Newbury Park, Calif.: Sage.

Foucault, Michel. 2001. *L'herméneutique du sujet. Cours au Collège de France (1981–1982)*. Paris: Gallimard/Seuil. (*The hermeneutics of the subject: Lectures at the Collège de France, 1981–1982*. New York: Picador, 2006).

Friedman, Batya. 1996. Value-sensitive design. *Interactions* November–December 1996: 17–23.

Fukuoka, Masanobu. 1985. *The natural way of farming: The theory and practice of green philosophy*. Trans. F.P. Metreaud. Japan Publications.

Gardiner, Stephen M. 2011. *A perfect moral storm: The ethical tragedy of climate change*. Oxford: Oxford University Press.

Gladwell, Malcolm. 2010. Small Change. *The New Yorker*, October 4, 2010. www.newyorker.com/magazine/2010/10/04/small-change-3.

Goujon, Philippe, and Bertrand Hériard Dubreuil (eds). 2001. *Technology and ethics: A European quest for responsible engineering*. Leuven: Peters.

Graeber, David. 2015. *The utopia of rules: On technology, stupidity and the secret joys of bureaucracy*. New York: Melville House.

Granovetter, Mark. 1973. The strength of weak ties. *American journal of sociology* 78(6): 1360–1380.

Gregg, Richard B. 1936. *The value of voluntary simplicity*. Wallingford, Pa.: Pendle Hill. www.soilandhealth.org/03sov/0304spiritpsych/030409simplicity/SimplicityFrame.html.

Hardin, Garrett. 1968. The tragedy of the commons. *Science* 162: 1243–1248. www.garretthardinsociety.org/articles/art_tragedy_of_the_commons.html.

Harvey, Peter D. 2000. *An introduction to Buddhist ethics: Foundations, values and issues*. Cambridge: Cambridge University Press.

Hershock, Peter D. 1999. *Reinventing the wheel: A Buddhist response to the information age*. Albany, N.Y.: State University of New York Press.

Hershock, Peter D. 2012. *Valuing diversity: Buddhist reflection on realizing a more equitable global future*. Albany, N.Y.: State University of New York Press.

Hershock, Peter D. 2014. *Public Zen, personal Zen: A Buddhist introduction*. Lanham, Md.: Rowman & Littlefield.

Herwig, Oliver. 2008. *Universal Design: Solutions for a barrier-free living*. Basel: Birkhäuser.

Himanen, Pekka. 2001. *The hacker ethic and the spirit of the information age*. New York: Random House.

Jamieson, Dale. 2014. *Reason in a dark time: Why the struggle against climate change failed – and what it means for our future*. Oxford: Oxford University Press.

King, Gary, Jennifer Pan, and Margaret E. Roberts. 2013. How censorship in China allows government criticism but silences collective expression. *American political science review*, May 2013. doi:10.1017/S0003055413000014.

Knappenberger, Brian. 2014. *The Internet's own boy: The story of Aaron Swartz* (documentary film, 105 mins). Luminant Media. https://archive.org/details/TheInternetsOwnBoyTheStoryOfAaronSwartz.

Kumarappa, Joseph Cornelius. 1948. *Economy of permanence*. www.scribd.com/doc/38457084/Economy-of-Permanence-Kumarappa.

Levy, Steven. 1984. *Hackers, heroes of the computer revolution*. New York: Dell Books.

Mair, Peter. 2013. *Ruling the void: The hollowing of Western democracy.* London; New York: Verso.

Meadows, Donella H., Dennis L. Meadows, Jorgen Randers, and William W. Behrens III. 1972. *The limits to growth: A report for the club of Rome's project on the predicament of mankind.* 2nd edn. New York: Universe Books.

Meijers, Anthonie (ed.). 2009. *Philosophy of technology and engineering sciences (Handbook of the philosophy of science, vol. 9).* Amsterdam: Elsevier.

Merkel, Jim. 2003. *Radical simplicity: Small footprints on a finite Earth.* Gabriola Island, B.C., Canada: New Society Publishers.

Mitcham, Carl. 1994. *Thinking through technology: The path between engineering and philosophy.* Chicago, Ill.: University of Chicago Press.

Mitcham, Carl. 1997. *Thinking ethics in technology: Hennebach lectures and papers, 1995–1996.* Golden, Colo.: Colorado School of Mines.

Ostrom, Elinor. 1990. *Governing the commons: The evolution of institutions for collective action.* Cambridge: Cambridge University Press.

Oudshoorn, Nelly, and Trevor Pinch (eds). 2003. *How users matter: The co-construction of users and technologies.* Cambridge, Mass.: MIT Press.

Puech, Michel. 2010. *Développement durable: un avenir à faire soi-même.* Paris: Le Pommier.

Rabhi, Pierre. 2007. *As in the heart, so in the Earth: Reversing the desertification of the soul and the soil.* Trans. J. Rowe. Rochester, Vt: Park Street Press.

Rifkin, Jeremy. 2009. *The empathic civilization: The race to global consciousness in a world in crisis.* Cambridge: Polity Press.

Schelling, Thomas. 1978. *Micromotives and macrobehaviors.* New York: Norton.

Schumacher, Ernst Friedrich. 1973. *Small is beautiful: Economics as if people mattered.* New York: Harper and Row.

Shirky, Clay. 2008. *Here comes everybody: The power of organizing without organizations.* New York: Penguin.

Singer, Peter. 2002. *One world: The ethics of globalization.* New Haven, Conn.: Yale University Press.

Slovic, Paul. 1987. Perception of risk. *Science* 236(4799): 280–285.

Stiglitz, Joseph E. 2012. *The price of inequality: How today's divided society endangers our future.* New York: Norton.

Surowiecki, James. 2004. *The wisdom of crowds.* New York: Anchor Books.

Tainter, Joseph A. 1988. *The collapse of complex societies.* Cambridge: Cambridge University Press.

Tapscott, Don, and Anthony D. Williams. 2006. *Wikinomics: How mass collaboration changes everything.* New York: Portfolio Hardcover.

Tavani, Herman T. 2004. *Ethics and technology: Ethical issues in an age of information and communication technology.* 3d edn. Hoboken, N.J.: John Wiley & Sons.

Thaler, Richard H., and Cass R. Sunstein. 2008. *Nudge: Improving decisions about health, wealth, and happiness.* New Haven, Conn.: Yale University Press.

Verbeek, Peter-Paul. 2011. *Moralizing technology: Understanding and designing the morality of things.* Chicago, Ill.: University of Chicago Press.

Verbeek, Peter-Paul, and Adriaan F.L. Slob (eds). 2011. *User behavior and technology development: Shaping sustainable relations between consumers and technologies.* New York: Springer.

von Hippel, Eric. 2005. *Democratizing innovation.* Cambridge, Mass.: MIT Press. http://web.mit.edu/evhippel/www/democ1.htm.

Watsuji, Tetsurō. 1996. *Rinrigaku (1937–1949): Ethics in Japan.* Trans. Yamamoto Seisaku, and R.E. Carter. Albany, N.Y.: State University of New York Press.

Weinberger, David. 2007. *Everything is miscellaneous: The power of the new digital disorder.* New York: Times Books/Henry Holt.

Wellman, Barry. 2001. The rise (and possible fall) of networked individualism. *Connections* 34: 3–32. www.insna.org/PDF/Connections/v24/2001_I-3-4.pdf.

Winner, Langdon. 1977. *Autonomous technology: Technics-out-of-control as a theme in political thought.* Cambridge, Mass.: MIT Press.

Wu, Tim. 2011. *The master switch: The rise and fall of information empires.* New York: Alfred A. Knopf.

Zittrain, Jonathan. 2009. *The future of the Internet: And how to stop it.* New Haven, Conn.: Yale University Press.

6   Ordinary Wisdom in the Technosphere

6.1 Which Wisdom

Wisdom, but Practical

Why wisdom and why is it an end in itself? This question has been amply addressed in the previous chapters from a general point of view. The answer is that wisdom, for a self, concerns being a self, that is to say being, and being is an end in itself for a sentient being. This immanent value applies to all living and sentient beings and even more to conscious beings.

Practical wisdom in the technosphere is modern *phronēsis* in the lineage of Aristotelian ethics. Aristotle is revived in numerous ways by modern scholarship, most of the time with a central or partial interest for his pragmatic doctrine of practical wisdom, *phronēsis* in Greek. However, there is still much to do before the elaborated judgment skills called *phronēsis* gain any influence in the field of real action (at work, at home, in a shop, online, etc.). Ordinary technoethics aims at such results. I will take Reeve (1992) as a guide through Aristotle's "practices of reason" in ethics. Reeve (2013) offers a precious working tool for examining Aristotle's texts on practical wisdom in the *Nicomachean Ethics*. More general studies about wisdom include important discussions of *phronēsis* (Sternberg 1990; Cooper 2012). Neo-Aristotelian ethics puts to work the concept of *phronēsis* or its derivatives (Hursthouse 1999; Bynum 2006; and in a sense Korsgaard 2008). Comparative Aristotelian and Confucian ethics studies (Angle and Slote 2013) are a major inspiration for my version of practical wisdom.

Phronēsis can be considered the archetype of a new technoethical wisdom, first because it is defined by its difference from theoretical wisdom, *sophia*, the "noble" form of wisdom that belongs to knowledge, to science, not to the realm of action. *Sophia* is the higher subject for philosophy (*philo-sophia*, the love and search for *sophia*). Some authoritative scholars like Aubenque (1963) insist that *phronēsis* is a circumstantial judgment, a non-universal judgment requiring specific training and depending on a multitude of contingent features in every specific occasion (*kairos*). Reeve describes an even more ambiguous situation:

First, *phronēsis* is more a kind of perception than it is a kind of knowledge of universals. Second, the kind of perception it is none the less
crucially involves knowledge of universals and of how to bring them
to bear appropriately in particular situations.

(Reeve 1992: 72–73)

In the Western source of prudential and sapiential ethics, Aristotle, the link
to knowledge is then ambiguous. This link, even if it is not one of deduc-
tion or mechanical application, still entails representative knowledge, the
typical ontological and epistemological structure of Western civilization.
With representative knowledge comes the entire ontology of modern technology, largely originating in Aristotle's philosophy (Heidegger 1939).
Doctrines of prudential and sapiential behavior remain a rich and complex
philosophical source but there is this "cognitive" or "representational"
limit to the Aristotelian renewal in ethics. The supplement I suggest for
going beyond this limit is a radical dewesternization of virtue ethics and
wisdom ethics. From Zen and Eastern inspirations, it is possible to concep-
tualize a technoethics of the ordinary that starts neither from meta-ethical
knowledge nor from doctrinal norms but simply offers a set of wise practices for a self-constituting person. Only a wisdom approach to technoethics could match the specifications of contemporary technoethics: being
normative and pluralist at the same time. Wisdom interpreted as the constitution of a self in the particular context of the technosphere allows defining norms while insisting on the personal elaboration of these norms.
Phronēsis then is not only finely tuned to the circumstances but also finely
tuned to and by the self. The whole of ethics can be seen as a quest for
wisdom that includes "value experiments" in the process of shaping oneself
as a self – technoethics draws this from recent wisdom studies (Kane
2010).

Table 1.4 advanced a basic description of wisdom in six practices,
which have already been put to work in the previous chapters: awareness,
autonomy, harmony, humility, benevolence, and courage. As an extended
"proof of concept," in the engineering sense of the term, I will offer a
series of technoethical applications of wisdom practices throughout this
chapter. This task requires a more substantial and specified reference list.
In going through this list (Table 6.1) I will not flatly lay out specific technoethical applications, case studies, and concrete examples as bits and
chunks of "what to do." Instead, I intend to use them as instances of the
type of action and of the *level of action* that are specific to ordinary technoethics in my approach. Table 6.1 is a tentative list of the meanings and
practices related to wisdom that could prompt wise personal action in
modernity.

In a first step, Table 1.4 only mentioned the *fundamental practices* of
wisdom, which was enough to give substance to the notion and to support
the references to wisdom in the previous chapters. For the present step, we
need to deploy the *practical virtues and skills* of wisdom, creating a more

Table 6.1 Wisdom Reference

Wisdom fundamental practices
Awareness
Autonomy
Harmony
Humility
Benevolence
Courage

Wisdom practical virtues and skills
Desirable acquired states
   Serenity
   Authenticity
   Consistence

Desirable acquired capacities
   Integrity (Honesty)
   Generosity
   Resistance

Necessary sensitivities
   Compassion (sensitivity to the suffering of others)
   Empathy (including for positive feelings)
   Aversion to violence

Ethical skills to be practiced and improved
   Non-confrontation
   Non-violence
   Frugality (temperance)

Useful practices
   Meditation
   Life hygiene (including diet and the management of addictive substances)
   Bodily maintenance through exercise (from daily exercise to martial arts,
      without any "sport" meant as competition)

---

detailed and more concrete list. Every item in this table is destined to work in combination with other items and in harmony with them. Every item is connected to numerous others in its very definition. Section 6.3 below will elaborate on the set of "wisdom fundamental practices," one by one, and section 6.4 will then derive the "wisdom practical virtues and skills" from this table.

The absence of some notions that are usually active in ethics may be disappointing in this table. I argue that they do not belong to wisdom. It does not imply that these virtues are less important or less valuable in themselves. Wisdom will *apply to* questions concerning happiness or friendship, for example, but the questions of happiness and friendship are not components of wisdom as I take it. To found a new technoethics, the ethical aspect of self-construction necessitates wisdom as a well-identified and independent, though richly diversified, pursuit.

Applying Zen-like wisdom to modernity has already given birth to interesting endeavors and the synthesis I am presenting here takes them into account. Some of this research, like Wei-Hsun Fu and Wawrytko (1991), elaborate on the links between contemporary examples and classic references in Buddhism, without much philosophical analysis, relying simply on resemblance and analogy – in spite of the authors' solid Buddhist scholarship. Some others, like Gordhamer (2008), are well aware of the cultures of modernity, geek and managerial culture at least, and they formulate stirring slogans, but in the style of gurus without solid philosophical ground.

More forceful predecessors are the Transcendentalist philosophers: Emerson and Thoreau. They tried to offer wisdom ethics as the response to modernity's challenges in correspondence with the project of this book. They offered a method and some instances of applied wisdom in a modern context, integrating Hellenistic schools of wisdom and Asian philosophy. To give an idea of the method that led to the Wisdom Reference Table below, see Cafaro's analysis of the virtues that are mentioned in Thoreau's *Walden* (Cafaro 2004: 55–56). Its extensive catalog of 105 virtues and skills largely overlaps Table 6.1. Cafaro offers a synthesis which leads to the following categories: Personal virtues/Social virtues/Intellectual virtues/Aesthetic virtues/Physical virtues/Superlative virtues (Cafaro 2004: 57–58). The synthesis itself evolves into this short list: "freedom, self-reliance, personal focus, self-knowledge, connection to nature, and philosophical reflectiveness" (Cafaro 2004: 117). Concerning Emerson's ethics, equivalent catalogs are implicitly employed by Robinson (1993), Van Cromphout (1999), as well as other classic studies on the ethics of the Transcendentalists. Table 6.1 is a structuring hypothesis that intends to update the Transcendentalists' notion of modern wisdom with a stronger reference to Buddhism, Zen and Ancient Chinese ethics and with significant input from modern virtue ethics and wisdom research.

An Alternative Framework for Modernity

Wisdom provides an alternative ethical framework for modernity. The request for change is clearly articulated in philosophy of technology: "a fruitful reform of technology must be one *of* the device paradigm and cannot allow itself to be confined *within* the framework of technology" (Borgmann 1984: 155). Technology needs no technological optimization, no technological fix, no better engineering. The "countercultural function" (Hershock 2005: 28) of Buddhism and Zen doctrines is a valuable resource for modernity. Through an adapted set of virtues and skills, a non-conventional system of values can be applied to the technosphere: a radically non-technological approach to technology emerges, focused on the personal practice of wisdom. There is a paradox in the first moment of this enterprise because the *Asian* traditions that support these values do not interpret "action" and "practice" in the Western mechanical and teleological sense. The most universal and deepest inspiration on this point is Lao Zi, the Daoist master: the right action is for him non-action (*wu wei*). Non-action is how nature performs her accomplishments (Lao Zi 500 BCE:

Sections 48, 63); it is how to flourish for every being. More directly it is the Way (*dao*). Lao Zi taught us in advance that superior virtue is not a planned goal that must be achieved (Lao Zi 500 BCE: Section 38); virtue does not apply an engineering plan to reality.

Wu wei concerns the self before concerning the world. Slingerland (2003) analyzes *wu wei* as "effortless action" and essentially a paradox. Although the paradox remains, *wu wei* runs through classical Chinese philosophy as a structuring metaphor. One of its forms is particularly significant: when *wu wei* generates a "technology of the self" in a process of "inner training" (Slingerland 2003: Chapter 4). In a later book Slingerland insists on spontaneity in *wu wei* (Slingerland 2015). Spontaneity is a typical wisdom skill that connects with the Buddhist "virtuosity" in Hershock and connects in Slingerland's work itself with the neurological approach to natural ethical skills in Varela (see Section 4.2). The paradox in *wu wei* dewesternizes self-construction in preventing mechanical self-engineering.

But *wu wei* is also a paradoxical method for transforming the world. Chinese strategists conceived the right action as a complex virtuosity relying on the potentials of a situation and not on a previous modelization of this situation. The art is to act *as little as possible* so that the situation evolves by itself into the wished configuration, following as much as possible its own internal potentials. This method is exactly the opposite of the engineering mind in the West, which favors the most powerful action possible for dominating nature and imposing the wished configuration on reality. The sage does not figure out a complex mechanism of cascading means and ends that will mechanically deliver by force the expected outcome, all of this being planned from a perfect knowledge of the situation. On the contrary, the sage uses the awareness of the global situation to intervene as little as possible, by what I call a microaction or a series of consistent microactions – and even when possible by non-action. Ideally, the Chinese general would scratch the sand with his cane, not to draw a plan, but to divert an almost invisible trickle of water, so that it flows into this tiny stream rather than into that one, and the tiny stream flows into a slightly larger one, all the way down to the river, adding just the supplement of water that is needed to overwhelm one little dam on the river, then a larger one on the next river, and in the end a massive flooding drowns the enemy camp which was set up on the shore of the large river below (Jullien 1997). The general is typically *victorious without fighting*.

This strategy must not be understood as a superior means-and-ends course of action; that would be engineering and technological method. The turning point is in relinquishing the command-and-control meta-attitude, which is the essence of the Western vision of action. Hershock's argument (Hershock et al. 2003) pleads for a "turning away from technotopia" which essentially resides in a turning away from this command-and-control meta-attitude. He gives the reason for this change: "A life centered on the values of technologically achieved autonomy and control commits us to an intensifying cycle of perceived poverty and fleeting satisfaction" (Hershock et al. 2003: 595). Hershock explains the logic of this cycle, which is the paradox of control: "Thus, the better we get at controlling our circumstances, the more we will find ourselves in circumstances open to and requiring control" (Hershock et al. 2003: 595). What is depleted in this meta-attitude is "our attentive resources – our capacity for meaningful and shared mutual attunement and contribution" (Hershock et al. 2003: 597). The challenge for an alternative ethical framework in modernity is in overcoming the basic meta-attitude of command-and-control. My perception is that the only method for such a radical shift is to reconceive human presence in the technosphere on the level of the individual person (and even on the level of her ordinary practices in the technosphere).

The wisdom evoked here is not a *telos*, an end in itself, a perfect state that is a goal to be reached by appropriate means and then rested content with. It is a process of self-construction that includes doubt and uncertainty and first of all learning from mistakes. In the world of material and informational abundance, for instance, the most practical and realist method for frugality is to check whether one does not need or want this precise item, material or immaterial, by candidly trying it and thereafter authentically assessing this experience.

In this sense, technoethical wisdom is the continuing effort of philosophy in its most original nature, philosophy as a way of life. This meaning of philosophy has been lost through a strange and contingent historical evolution, suggests Cooper (2012). Maxwell (2007) is more aggressive with the academy. I would prefer non-confrontation and humility, plus an orientation toward ordinary individual practice. A *Journal for Modern Wisdom* has been launched, trying to restore, says its manifesto, "the ability to be reflective or meditative about one's priorities in life, and to cultivate a positive self-image in doing so" (Irvine 2011: 8). This project is closer to my notion of wisdom in self-construction, especially in its ordinary accomplishments:

> Whether through calm introspection, artistic expression, or enlivening activities like jogging on a winter's morning or dancing beneath the stars – what's achieved is that highly self-aware state of generosity and openness that's variously called conscientiousness, enlightenment, or wisdom.
>
> (Irvine 2011: 8)

Behind these inspirations lies John Dewey's progressive conception of philosophy. It is centered on education and pragmatic action, not directly on wisdom. Self-education and its application to actual behaviors, as explained in Dewey's work, still provide significant material for a wisdom ethics of modernity. His basic intuition calls for "reconstruction" in philosophy (Dewey 1920). Dewey's renewal of individualism in his 1929 articles for the *New Republic* (Dewey [1930] 1984) gives a pragmatic version of the self-construction background of modern wisdom. This version can be used to counterbalance the mystical tendency in Daoist and Zen versions. For Dewey, education and self-education give the method for breaking the stranglehold of modern moral impoverishment in a typically technoethical way, through the potentials of the technosphere:

> For machinery means an undreamed-of reservoir of power. If we have harnessed this power to the dollar rather than to the liberation and enrichment of human life, it is because we have been content to stay within the bounds of traditional aims and values although we are in possession of a revolutionary transforming instrument.
>
> (Dewey [1930] 1984: 87)

However, most of the formulations of the modern predicament in terms of wisdom remain, intellectually, in a means-and-ends technical mechanism. In more or less directly, then, favoring power as the necessary means for action, they interpret power as political domination and technological engineering of reality (Table 1.3), missing the genuine power of wisdom: power over oneself, the neglected resource.

Concretely, in some versions of modern wisdom which at first sight are in perfect harmony with my argument, such as Kekes (1995) or Spence (2011a, 2011b), the Western style of analytical decomposition plus (moral) engineering prescription is so pregnant that in the end such research is at risk of falling back into sterile scholasticism. Moral wisdom means essentially increasing the *control* of persons over their actions, says Kekes repeatedly, in a way that belongs to the command-and-control paradigm of Western technology. Spence's eudaimonic model often sounds like a moral engineering specification procedure.

The opposite approach, *yang sheng*, to "feed one's life," in Chinese philosophy, does not lead to a eudemonic system (a technique) of happiness or to a moral technique (optimization by calculus) as in utilitarianism. Jullien (2005) shows that this original self-construction process tries to seize the original intensity of life, avoiding the teleological (technological) obsession that prejudicially defines happiness as the end. A living capacity for detecting and exploiting harmony in every situation and in every moment leads to a different practice. The conduct of one's life does not depend on predefined truth, ontology, or theology: therefore the construction of oneself is not an engineering process, acting according to plan. It requires something totally different, a direct connection of the sage's personal development with cosmic harmony, and this goes with a total contempt for the alleged necessary intermediaries, which are political or ideological systems. Jullien's analysis helps understand why genuine wisdom doctrines have been repressed by all kinds of temporal and intellectual powers, by all kinds of institutions. Wisdom and the sage "grow" as life itself, according to the most ancient Daoist paradigm, in a splendid independence. Independence gained in a superior understanding of interdependence is a *wu wei* definition of modern sagehood.

After this first internal condition – the self-construction background of wisdom – a second condition, external, must obtain for technoethical wisdom: an acceptably open political and economic environment. The presumed situation of relative economic welfare and acceptable democracy is the real situation of a large part of mankind in the twenty-first century and it is the desired situation of the majority of those that are for the moment excluded from it. Technoethics is about the pursuit of wisdom in this context. Wisdom is always in context. The context for technoethics in the twenty-first century is then philosophically specific. Market economy under legislative regulation and elective democracy with a balance of institutional power and freedom of speech: these questions were rightly at the center of attention when the common task was to bring these realities into existence. In the times and places where these conditions now obtain, this common concern must yield priority to the concern of the self, the neglected question in modernity. Of course, it remains an issue of prime importance to preserve the social framework in the background and even to enlarge and improve it, in terms of justice in particular. In these matters, what is required from the self is citizen virtue, most of the time ordinary: staying well informed, protecting the collective capacity for critical assessment, preserving some time and energy for taking care of the commons.

A resolute optimism is at work in the project of technoethics as wisdom, particularly concerning the external context. More than optimism, I underline the *constructive* mind, which calls for action and is eager to realize the promising potentials in the technosphere as it is. This optimism assumes a *candid* attitude, not a naive one. The sage knows that things are not so simple but action requires resoluteness. Hence a candid meta-attitude, which falls under the wisdom practice of courage (Table 6.1).

6.2 Ordinary Engagement

Engagement as a Value and a Wisdom Value

The specific concept of engagement that allows a technoethics of the ordinary needs a non-Western immanentism of engagement. It can be sketched essentially from Daoism and Zen. This immanentism neutralizes the risk of a narrow ideological commitment in the notion of engagement. Ideology would lead to a misinterpretation in terms of dominance, the wrong sort of power.

The richest ancient source for Daoism is Zhuang Zi (475 BCE), whose importance goes far beyond his belonging to this philosophical orientation. Through numerous concrete formulations, coherent with the pragmatic tone of Chinese culture, the Dao appears in Zhuang Zi not only as the Way in the cosmological and somewhat mystical sense that it had in Lao Zi but as a practical way for human conduct. Practicing the way of the Dao is somewhat of a pleonasm, in fact, because the Dao is the nature of everything and human action has no other way than entering into its harmony. Following the Dao is also somewhat of a paradox because this action is in a sense non-action: it is the entering into the harmony of the global and not the egoistic perturbation of this harmony. What is surprisingly modern in Zhuang Zi is the presence of a substantial ethical self, a human person who takes her own conduct in charge. The simplistic assertion that there can be no "subject" in Asian philosophy does not hold in the face of the texts. The "way" means the pattern of one's practice, an active pattern, which relies on a correct understanding of the Dao. However, nothing follows as the simple consequence of a priori understanding. The ethical immanentism in Ancient Chinese philosophy deserves special attention as it helps formulate an immanent ethics of engagement: asking questions and thinking one's way in the very process of walking it (*caminar preguntando* in the Spanish of South American alternative philosophy).

American Transcendentalism offers another valuable resource for immanentism in ethics. I consider it in a wide acceptation including the educational and communitarian "utopian" projects in the Transcendentalist movement. Lifestyle was then conceived as the right level for authentic ethical reform. In lifestyle reform a real existential virtuosity takes shape. As is well known, Thoreau's foundation of modern wisdom begins with self-reliance, meaning engagement: "The reform which you talk about can be undertaken any morning before unbarring our doors. We need not call any convention" (Thoreau [1849] 1985: 102–103). According to Stoehr's survey of the Transcendentalist movement, a quasi-identification of thinking and acting was the most important and profound accomplishment of this movement (Stoehr 1979: 154). Stoehr links this orientation with Gandhi and Martin Luther King and with the new societal sensibility in the 1960s and 1970s, situating the Transcendentalists at the beginning of the ethical reform of modernity:

> They were among the first to confront the world as we know it – a world of "too much," in which too many possibilities offer themselves, too many careers, too many possessions and pleasures, too much complexity and ramification, too much leisure in which to think about all these alternatives, and to dream up more.
>
> (Stoehr 1979: 155)

The Transcendentalist solution to the modern ethical predicament is not a doctrinal solution, a recipe for personal or social transformation. Their solution consists in transmitting a feeling, through poetry and fiction as well as through intellectual insights. This feeling embodies the immanence of value in the self, connected with global harmony. Engagement is the step to put this ethical immanence in action.

Heidegger's philosophy must be recalled when making engagement the responsive attitude toward modernity. Heidegger's personal political and careerist engagement with Nazi Germany was a wrong choice, no doubt.

But there is much to learn in Heidegger's analysis of an engagement that can discontinue modern dereliction. Schürmann (1987) explores Heidegger's philosophy of action and especially of engagement. I follow him on the primacy of action in Heidegger, a primacy over ontology and any other sort of foundational thinking. Action is the condition for thinking, says Schürman (1987: 291). In his interpretation, Heidegger nullifies the difference between the question of what we ought to think and what we ought to do. What becomes possible from such a stance is what I will call after Hershock an *appreciative and responsive attitude* toward modernity. The relationship of Heidegger's fundamental conceptions with Daoism and Zen is established by solid scholarship (May 1989).

In the twentieth century, Arne Naess' philosophy of self-realization is another resource for a philosophy of engagement. As we have seen (Chapter 4), Naess' self-realization implies the concrete realization of harmony with the ecosphere (Naess 2005: Chapter 45). I suggest extending this engagement to the technosphere and the infosphere. In the typical non-ideological and non-confrontational style of Naess' philosophy, engagement is most of the time ordinary engagement: personal life reform as he did for himself and occasionally non-violent militancy.

The final step toward engagement as operational wisdom involves concentrating on microactions in order to reorient the vast array of possible engagement toward the precise practice of self-construction. Micro-engagements are a practice of wisdom that offers a non-ideological way for self-construction in the technosphere.

In an ethics of wisdom, values play a role that differs from their role in an ethics of norms. For this reason, without refusing the term "value," I prefer calling the elements of Table 6.1 "practices" of wisdom. Nevertheless, in order to offer an acceptable ethical engagement, technoethics must redeem values and norms in philosophy and fully endorse the human capacity for self-normed conduct. Values can be taken as constituents of reality only if we discard the Western ontology of material causation and objectivity, as Hershock (1999) suggests from a Buddhist point of view. Once free from the very narrow ontology of "objectivity," the philosopher can start afresh from the perception of the world by the self. Situations are perceived and understood through values because human situations are constituted by values. Therefore, the grip we have on these situations is due to values. The really active attitude is basically axiological and not technical, as it would be in front of an "objective" set of problems. Through the (ironically coined) "Buddhist technology" that he describes, Hershock (1999: Chapter 7) is led to the ordinary active meditation that I put forward in this chapter as a key practice of wisdom in the technosphere.

The adequate response to the challenge of the technosphere is not a system of control. Instead, it works like a response in a conversation. Abandoning values of control opens the way to the virtues of detachment and serenity in the accompaniment of reality through engaged action. For instance, engaged micro-decisions concerning transportation during an ordinary day cannot be blocked by a "realistic" or simply utilitarian analysis. This narrow approach misses the issue in its factual complexity because a myriad of factors are intertwined and knowledge about them is terribly imperfect. Engaged micro-decisions refer to tentative values, such as autonomy (pleading for a car ride), ecological harmony (pleading for public transportation), or frugality (pleading for reexamining the reasons to travel).

· The South American motto, *caminar preguntando*, means that one can walk while asking questions about the walk. It symbolizes the humility of wisdom. Engaged action is not the fulfillment of a program but rather a progressive moral education and self-education, which is scalable and involves different "stages" of moral development (Kohlberg 1981). This method is incompatible with a preexisting global set of values because it is the search for and not the simple implementation of these values. The sage is at the same time concretely engaged with values and in search of these values. Values, then, refer to a certain form of caring for the self – and most of the time in a sharing process with others where self-education and mutual education are interdependent. For the self as for the community, a form of moral maturation is at stake in the engagement with values.

The Ordinary as Engagement: Zen Meditation and Samu

The practical virtues of wisdom (Table 6.1) participate in ordinary engagement; first of all *awareness* because ordinary engagement is entirely made of mindfulness. No other component than extreme awareness characterizes its nature. Then *autonomy* follows because what has to be put aside is the ordinary disengagement in modern life: constant distraction and shallow preoccupation, all of them caused by constant solicitation (work environment, media, relatives, etc.). Reappropriating awareness and responsibility for ordinary life is a conquest, an accomplishment of self-reliance and autonomy. Then comes *harmony* and the intention of achieving the maximum possible harmony. This practice is self-supporting because immersion in a harmonized environment of small things makes it easier to remain concentrated on the depth and importance of the ordinary. *Humility* is involved in ordinary engagement because the time and efforts invested in small things do not bring the same narcissistic rewards as socially valued pursuits.

In philosophical and literary traditions, "engagement" naturally conveys an idea of the extraordinary more than of the ordinary. In French philosophy, "engagement" is the name for Jean-Paul Sartre's theory of political commitment. In most cultures, valued attitudes are extraordinary attitudes, such as fighting in a war or sacrificing one's life to a cause, which could include a domestic cause (an invalid parent for instance). We tend to value ideological contexts, where the grandeur is provided by reference to a transcendent valuation system. The dewesternization frame that I suggest reinvents engagement as an ordinary attitude.

The richest reference to ordinary engagement lies in the Buddhist doctrine of *samu*. Meditative engagement of the ordinary, taken as the core practice of a sapiential technoethics, is essentially an adaptation of *samu* to the technosphere. *Samu* (a Japanese word used in Zen schools) means the practice of meditation within ordinary activities. It was originally intended for monks retired in a monastery. The concerned activities were cooking, cleaning, and gardening; the ordinary chores of a monastic community based on frugality. Beyond this minimal set, there is no reason *samu* could not be extended to the whole set of ordinary activities in the technosphere. Cooking, cleaning, and equivalent basic activities still exist in a sophisticated form, but this is not a hindrance to meditation. If it is possible to meditatively sweep the floor of the monastery it must be possible to meditatively vacuum the apartment, and so on with the dishwasher, every washing or cooking device, and the lawn mower. A lot of the moments of life at work, and even more during transportation from home to work, can be moments of *samu*. Ives (1992: 36) goes as far as to translate *samu* by "work practice": there is no reason to limit it to monastery or household maintenance. With an important exception for the activities that distract our attention (mass media, commercials, and much of the entertainment industry), all activities in principle can be tried as a support for meditative practice.

The wisdom effort in *samu* is awareness or mindfulness, extending in two directions: awareness of the self and awareness of the interdependence of all things. Both these directions are essential for technoethics. The intention to harmonize, internally and externally, is the proper dynamic of *samu*. It is not exactly a contemplative state, and it belongs to the class of *engagements*.

Traditional sources on *samu* insist on the meditation aspect of this practice: surprisingly they are nonetheless full of suggestions for present-day technoethics. In early introductions of Japanese Zen culture to the West, one of them concerning martial arts and the other one concerning the way of tea, *samu* as meditative engagement in a "technique" was pivotal. In the preface to a source book on martial arts for the West by Herrigel (1953), D.T. Suzuki refers to the classical Zen author Mazu (Baso in Japanese, eighth century) to affirm that Zen is the "everyday mind" and this everyday mind is no more than "sleeping when tired, eating when hungry" (Herrigel 1953: viii). In the same way, the *Book of tea* was meant to open Western minds to Zen spirituality (Okakura 1906). It can be entirely read as a transposed version of technoethics. Concerning tea, flower arrangement (*ikebana*), the art of house decoration, the art of garden, and other material matters, Okakura's book gives a striking image of the difference between the Western lack of concern and care for the ordinary and the Japanese engagement in the beauty and harmony of every detail in ordinary activities.

The original doctrine of *samu* relies on one of the fundamental non-dualities in Buddhism: the non-duality between meditation and activity.

Musō Soseki (fourteenth century) expresses this point in the following manner:

> People meditating on the fundamental carry out their ordinary tasks and activities in the midst of meditation and carry out meditation in the midst of ordinary tasks and activities. There is no disparity between meditation and activity.
>
> (Musō [13–14th century] 1994: 53)

The modern reference on this subject is Thich Nhat Hanh. He gives numerous examples and precepts for meditation in the midst of ordinary activities (Nhat Hanh 2000). He starts with the most ordinary ones that have belonged to meditation techniques from the beginning: breathing, smiling, sitting, and walking. He calls "engaged" practice of Zen (Nhat Hanh 2005: 51) the meditative and active reappropriation of ordinary activities like cooking and eating, plus any kind of practice that leaves the mind free. Practical recipes are given to maintain a meditative state while driving a car, answering the phone, etc.

In Poceski's (2007) interpretation of Chan Buddhism (the Chinese origin of Zen) and of Mazu's doctrine in particular, "the state of *ordinary mind* is the goal of practice" (Poceski 2007: 183). Ordinary engagement can be the core practice of a wisdom that is therefore less and less a religion, especially in the Western sense of the term. No supernatural power or status is ever mentioned but rather ordinary subtle perfections "manifested in such prosaic acts as fetching water and carrying firewood" (Poceski 2007: 183). These subtle perfections were celebrated as "sublime ordinariness" by the poet Pang Yun (Poceski 2007: 184).

For a modern and ethical application of *samu*, Hershock is once again the best guide. "Chan awakening can be described as the realization of dramatic and liberating intimacy with and among all things" (Hershock 2005: 81). Meditative engagement in the ordinary is taken by Chan and Zen schools as a rightful spiritual discipline "in the midst of everyday circumstances" (Hershock 2005: 305). No activity is potentially outside the scope of meditative reappropriation because "no activities are outside of practice" (Hershock 2014: 182). Being active adds to the value of meditation: "Meditation in the midst of action is a million times better than meditation in stillness" (Hershock 2014: 210). I will develop (Section 6.4) the idea of "virtuosity" that emerges from *samu* in Hershock's interpretation: "tireless responsive virtuosity in the midst of any activity" (Hershock 2014: 219). "Responsive virtuosity" is another telling expression for sapiential ethics.

This approach to meditation substantiates the ethics of the ordinary in Emerson and Thoreau (Robinson 1993: Chapter 4; Cafaro 2004: Chapter 5). A balanced philosophy of life in technology is given by Emerson in some essays (notably "Experience" and "Works and Days"). Applying his principle "all things are moral" (Emerson 1983: 289), Emerson also celebrates the ordinary in the 1844 lecture, "New England Reformers":

"The reward of a thing well done is to have done it" (Emerson 1983: 608) and "that is ever the difference between the wise and the unwise: the latter wonders at what is unusual, the wise man wonders at the usual" (Emerson 1983: 609). Thoreau's "everyday sacramentality" (Cafaro 2004: 27) is perhaps the most original feature of practical wisdom in *Walden* (Thoreau [1854] 1985). In *Walden*, the numerous instances of ordinary activities (domestic and agricultural) are more than symbols of interior reform. The sacred value of the ordinary, as Thoreau rightly understood it from Asian sources, is to be taken at face value, not only as a symbol.

There is something ironic in the idea of an ethics for unimportant activities. This irony stems from the paradox at the heart of the culture of consumption in a society of abundance: we celebrate the ordinary – malls and supermarkets are places of ritual and cult. But rashly despising consumer goods and sacredness is a sterile confrontational attitude. Understanding consumption in its contemporary function leads further than easy moral condemnation. In the messiness of everyday affairs there can be an ethical strategy, running through microactions, micro-decisions, and micro-attitudes. The turning point of this technoethics is similar to a Zen revelation: the importance of the unimportant.

Meta-engagement

Wisdom practice is a meta-engagement in continuity with the definition of values as meta-attitudes – and meta-attitudes govern the process of self-construction. We have already drawn from this kind of approach in the theory of importance (Frankfurt) and in the theory of moral maturation (Kohlberg). Heidegger's *existentials*, the quasi-transcendental structures of human existence and action, are meta-capacities and meta-engagements, determining the status of authenticity or inauthenticity of the *Dasein* prior to any particular action or decision. Contemporary theories of applied wisdom confirm this interpretation of wisdom as a second-order virtue (Kekes 1995) or as a general pattern for virtues. Spence (2011a) comments on wisdom as essentially a meta-attitude, a meta-practice, in his research in order to apply wisdom to the infosphere.

It seems that the fragmented and dissociated technosphere of today is in demand of a meta-level to restore meaning and significance. There is a constant danger, however, if this meta-engagement refers to preexisting knowledge or values. Therefore, meta-engagement in wisdom must keep an unwavering Zen indifference to doctrines and particular norms. For this reason, it must be self-foundational. An unwavering resolution is needed again because there can be no regression toward meta-meta-engagements. Engagement as meta-attitude in Frankfurt's ethics actually avoided the risk by concentrating normative authority in the will, the individual will of the self, and not reason (Frankfurt 2006: 3). Meta-engagement chooses morality or wisdom as a whole and thus avoids the paradox of the (already moral?) motives to choose morality (Frankfurt 2006: 28). "Unless a person knows what he already cares about, therefore, he cannot determine what he has reason to care about" (Frankfurt 2006: 23). As in Zen wisdom, there can be no "content" of wisdom in the sense of a doctrine beyond the inaugural and uncompromising resolution for awareness and care.

Exercise

Like "practice," the term "exercise" means at the same time *training* and *action*. In a philosophy of the self-constructing self, this dual sense is not ambiguous but rather revealing: action, and particularly moral or sapiential action, is the exercise and practice of the self in the dual sense of training (improving oneself into being a self) and being active as it is required by the nature of the self – being a "practicing" self in the legal acceptation of the term, not a "retired" self.

The identification of the self with its own practice is expressed by the Buddhist doctrine that awakening or wisdom is not a goal, a future experience toward which one strives. Instead, awakening and wisdom are practices, here and now, an effort and an engagement of the self in every present moment. As Hershock puts it for the modern reader: "In short, practicing Buddhism is not about *getting* enlightened, it is about *demonstrating* enlightenment" (Hershock 2014: 29). Thoreau's ethics clarifies the consequence of this "demonstration" in personal life with this provocative demand in the essay "Reform and the reformers": "I ask of all Reformers (...) to carry around with them each a small specimen of his own manufactures" (Thoreau [1846–1848] 2004: 184), that is to say: oneself. This demand contrasts with the indefinite setting of distant goals and the incessant renewal of promises that prevail in politics, religion, ideology, and sterile scholasticism. Wisdom imposes achievement not by being a goal but by being a practice. The sage has to personally exemplify awakening, not instruct it, says Zen philosophy (Hershock 2014: 151). The revolutionary ethics of this immanent engagement gives birth to a renewed ethics of modernity, relying on the actual transformation of the actor from the inside.

The Transcendentalists claimed that real engagement is near and daily, not in the self-promotion of moral rhetoric. Emerson makes it clear in a striking passage of the essay "Self-reliance" (Emerson [1841] 1983: 262) and all throughout the pages of *Society and solitude* (Emerson 1870). Thoreau's version takes a political turn in the famous *Civil Resistance* manifesto:

> The fate of the country does not depend on how you vote at the polls – the worst man is as strong as the best at that game; it does not depend on what kind of paper you drop into the ballot-box once a year, but of what kind of man you drop from your chamber into the street every morning.
>
> (Thoreau [1849] 2004: 104)

The lesson of wisdom that is transmitted to us by Buddhism and the Stoics, directly or through modern interpreters like the Transcendentalists, must be secured from a "perfectionist" interpretation. It is a demand for immanent engagement, meaning for us engagement in the ordinary of the technosphere. On the contrary, according to an idealist ethics of virtue, virtue is an ideal goal: a perfect state that the self reaches in this life or possibly a future life through a progressive improvement process: a sort of self-engineering. Concentrating on immanent exercise, in the humility of *samu* (ordinary activities taken as meditation practice), hopefully leaves no place for this planning of one's life according to transcendent standards.

6.3 Practices of Wisdom in the Ordinary Technosphere

The form of wisdom answering the challenges of contemporary technoethics was hypothetically presented in a short list of wisdom virtues at the beginning of this book (Table 1.4). In the advancement of the argument, these virtues proved to be practices (dynamic engagements) more than static virtues (teleological perfections). In this Section 6.3, I will develop the applicative dimension of these practices.

The practicing sage in the technosphere may appear under very different guises. Therefore, various options are available to instantiate each of the fundamental practices of technoethical wisdom. Its normativity is a sort of meta-normativity: it prescribes what are the questions to ask oneself and where are the crossroads to be approached with full conscience; but it does not prescribe which way to take. For instance, understanding the meaning and the potentials of frugality is a wisdom practice that may lead to possessing zero, one, or n cars; the value of n is not essential. More exactly, it is essential for the self but cannot be prescribed as a norm.

I want to emphasize two features in the following developments: how coherent the practices of wisdom are and how applicable to the ordinary of the technosphere they are. Coherence means that each practice of wisdom forms a system with the other practices. Imagining a "conflict of duties" between them remains possible but requires specially designed thought experiments, most of the time. Applicability means that the list of ordinary technology practices (Table 1.2) can be browsed from the point of view of each wisdom practice and generate an infinite number of possible instantiations. This contributes to ascertaining the "proof of concept" for a technoethics oriented toward ordinary wisdom. I will now follow Table 1.4, which is the first part of Table 6.1 ("Wisdom Reference").

Awareness: The Beginning of All Wisdom

Section 3.3 above has presented awareness as the ordinary virtue par excellence, in the context of the ethical significance of the ordinary in the present technosphere. The wisdom value of awareness brings depth and width to modern life. Ethical awareness replaces the traditional notion of ethical excellence, a notion that was dependent on the Western and "technical" values of perfection (the end for which means are to be found and deployed) and competition (a scale of accomplishment and merit). Being aware does not enroll one in a competition against the others or even a competition with oneself, to reach a perfect state of pure awareness. That final state was in the religious message of Buddhism, I admit, but now we are looking for disseminated ordinary micro-awakenings and not The Awakening.

Humility and autonomy are necessarily associated with awareness. Foucault's interpretation of Hellenistic moral doctrines alludes to a modest practice of the virtues, even if in ancient schools they were opportunities for moral arrogance. "What I had been acculturated into finding trite and windy edification Foucault made fresh and challenging," says A.A. Long about Seneca's moral prescriptions (in Bartsch and Wray 2009: 221). In ordinary technoethics, particularly, there is nothing "preachy" in the call for awareness. Moral preaching is an attempt at moral domination. On the contrary, sapiential awareness is the effort to liberate the self from the moral dereliction of today: ethical unawareness and irresponsibility, the absence of ethical care. A Zen-like concentration of attention gives the opportunity to "decolonize" modern consciences:

> The first real step toward realizing post-market economic and development realities, then, is to become mindful of the forces that presently define – literally shape and set limits to – how we are attentive.
>
> (Hershock 2006: 4)

Shunning mass media influence on our conscience, which is the main cause of the "colonization of consciousness," is then a priority (Hershock 2006: Chapter 4). But the meditative engagement in the ordinary brings positive value in a vast range of activities, beyond the liberation from the "alienation" of attention. Existential reappropriation begins with the reconquest of attention, in awareness – "If our attention isn't literally our own, what is?" (Hershock 1999: 82).

An imperative in the samurai's code was: "Live being true to the single purpose of the moment" (Yamamoto [1716] 1979: 68). It proceeds from the moral warning: "Negligence is an extreme thing" (Yamamoto [1716] 1979: 17). In the samurai's ethos, which is still approachable for us through martial arts, there are two dimensions of awareness and both can be applied to ordinary activity. One is perceiving the depth of the existential feeling in the moment of action, a sort of ontological engagement – "If one fully understands the present moment, there will be nothing else to do, and nothing else to pursue" (Yamamoto [1716] 1979: 68). The other one is perceiving the width of the present situation: the capacity to hold as a single reality one's body, the opponent's body, the ground and physical environment, discerning the harmonies and potentials of the situation as a whole. Then the right posture and the right move emerge naturally and they are accomplished without mental or physical effort. The usual mistake is to let one's mind wander everywhere, in anticipations, expectations, memories, fears, and hopes. This flow prevents fully inhabiting and exploiting the present moment. Awareness and concentration can be sustained in a continuing existential meditation. In the eighth century, Shantideva attributed "Far-Reaching Discriminating Awareness" to the sage as the practice of *vigilance*, an indispensable virtue (Shantideva 763: Chapter 9). We could say that a permanent *existential vigilance* belongs to the practice of wisdom in any sphere.

Thich Nhat Hanh gave this formulation of the core precept of Zen: "We should be aware. The most basic precept of all is to be aware of what we do, what we are, each minute. Every other precept will follow from that" (Nhat Hanh 2005: 70). He then displays its application in ordinary awareness experiences, reaching as far as "mindful consuming" in a consumer society (Nhat Hanh 2005: 94). We have seen the Buddhist origin of this practice with the notion of *samu*.

Let us start with awareness of speed and of the laws of physics concerning inertia and collision when driving a car or when being the passenger of a speeding car. An easy mental exercise can restore this awareness, periodically, while travelling. What if a tire, brake, or steering-wheel malfunctions now? What if on the other side of that curve a drunk driver speeds on the wrong side of the road? The idea is not to constantly imagine a crash, but it is to emerge from the false apprehension of the car as a set of armchairs and of the moving landscape as a kind of visual show displayed on the windows. Just being aware that one is a mass of flesh, trapped in a huge mass of steel, launched at a speed that exceeds the speed of the fastest runner: this mental experience is enough for a non-suicidal person to drive safely.

At a larger scale, awareness of the fossil fuel issue in the present world can change a lot in ordinary behaviors and then bring about major global consequences. Awareness of the fossil fuel chain of supply means awareness of its political consequences – which are: financing dictatorships that finance international religious terror – and of its economic and ecological consequences – which are: maintaining the power of oil corporations above any national or international authority. Every drop of fossil fuel used or non-used is a political microaction, from the point of view of ordinary awareness. It does not mean a moral sin, nor a political crime, but simply a question of awareness: the ultimate reason for international affairs to carry on as they do, in spite of virtuous institutional initiatives and intensive discourse production, is that the microactions of the end-actors are not coherent. The reasons for this ethical discrepancy point to the lack of awareness in individual actors. The lack of awareness in the ordinary actions and decisions in the technosphere is the real ideological blockage of coevolution. This is why an adapted form of Buddhism may help, through the arising, strengthening, and educating of awareness. It differs from an ecological or political "consciousness" which would arise from social critique and then apply top-down. Wisdom awareness works bottom-up. It starts in the meditative reappropriation of the most ordinary actions, transportation and heating or air conditioning, for instance.

In the infosphere, awareness of the traces that are left by any Internet and telephone activity is of central importance in technoethics. We know that secret services and marketing departments spy on us. What would be an acceptable level of digital transparency? Being aware that every search on Google produces data that are actively used by some firms and government agencies changes the current microactions but does not necessarily suppress it. For instance, one can decide that normal Google is usable for every search that one accepts as possibly public and that a secured anonymized connection is required in case something must remain strictly private.

The perimeter of secrecy depends on the level of toleration in the social environment: in some environments, a love affair is taboo; in some others, a same-sex love affair is a social shame; in some social environments political or religious opinions can endanger one's life; in some environments, certain diseases or the exact amount of money one earns are taboo. The level of accepted transparency is a social variable too and it can change, accompanying technology. Changing this kind of social variables can even be the purpose of personal and collective action, particularly if this dynamic awareness is invested in consistent microactions.

When one pays anything with a payment card, one must know that possibly anyone can be aware of this purchase. There is a moral problem if the customer is not aware, but since one is aware, the adaptation of one's behavior to this fact is a dramatically different moral problem. From a radical option (never using a payment card) to the opposite radical option (endorsing total transparency), there is an open range of technoethical choices. The lack of awareness implies here the absence of self-construction: living as an object in commercial and societal networks, not as a self. Technoethics prescribes that we first of all take full awareness of the plain fact that our bank uses and sells to other corporations all the details of our accounts and shopping lists. This is awareness, the first step. Then, technoethics fosters the capacity of self-reliant persons to act consistently, whatever their option is. Autonomy and courage will be needed; the wisdom practices of resistance, frugality, non-confrontation, and others (Section 6.4) will help. The issue of privacy begins with awareness and continues with consistence. This method does not deny the conspiracies and wars suurounding Big Data: it just deals with the pragmatic response.

The fully aware experience of a hot shower should be revered; it can be practiced as a kind of meditative routine in ordinary life. It includes a mindful thought for energy and water consumption, with all their economic and ecological consequences, which can be accepted under certain conditions (if not, a dramatic decrease in body hygiene is an option). Taking a mindful shower is a privileged experience of the comfort and reassurance that we may legitimately care for and accept to be attached to.

When a trip into the wild (hopefully) or some misfortune (sadly) deprives us momentarily of this experience, we understand its importance. An awareness exercise in wisdom is to experience this every day.

The practice of *samu* (Section 6.2), meditative awareness in every moment and in every action, is the base for taking responsibility and possibly taking action in the technosphere as it exists, with its boundless potential, on the one side, and its rampant disengagement, on the other side.

Yet, awareness is not the last word of technoethics because it is a state of mind and not a full-blown activity. Awareness is not the whole of the practical engagement that sapiential practice must be. But awareness is the beginning of all wisdom, without which no other practice or virtue would be a self-constructing practice.

Autonomy: Waging Invisible Moral Wars

Moral autonomy is vital for the sage in every epoch. The fact that the capacity to grow or supply one's own food has been lost is an intriguing characteristic of the present technosphere. The "bobos" and utopian militants of the 1970s reform movements have a point in signaling that this is the loss of something essential in human experience. Once again, awareness is the first step of wisdom: realizing that the majority of us could not survive without commercial food supply. It gives a different context to "consumer society" when we are reminded that we owe it our life, day after day. The next step after this awareness can be to restore more autonomy, first by becoming able to prepare one's food from more raw material (rather than processed food) and then by finding a supply chain for this raw material. Direct supply from a local farmer is still possible in rural areas. In urban areas, the hype for gardening (including on the roof) may be a lasting trend, and more realistically community associations appear everywhere in large cities for buying from local farmers and distributing among members independently of the commercial retail network.

Growing one's food, even in very small proportions, and being able to nourish others, family and friends, is a unique experience and practice – Zen-like or Walden-like. This experience is of course symbolic but it embeds a deeper genuine moral value: the value of autonomy. Here again, the microactions of individuals change the global picture. The governmentality (Foucault's notion) of people who are able to nourish themselves vastly differs from the present governmentality of people who would starve at the moment the retail chain breaks down.

The essential autonomy remains moral autonomy. We have seen that in recent virtue ethics and care ethics the conceptions of moral autonomy tend to focus on *importance*. In teenage angst and the subsequent crisis what becomes more and more unacceptable to teenagers is the impossibility to master their own attribution of importance (chatting online is more important to them than going to bed at a reasonable time, for instance, or going to a party is more important than visiting family relatives). What is unacceptable in the workplace is the bossy manager who dictates what must be done, when and how, in every detail, leaving no freedom to choose one's priority. The human person is hurt in her dignity when moral autonomy is denied to her, at any scale and in any context. "Wisdom is an understanding of what is important, where this understanding informs a (wise) person's thought and action," writes Nozick (1989: 267). In the present material and intellectual context, the practices and virtues of ordinary autonomy are largely a defense strategy because behind the appearance of an individualistic society, autonomy has its enemies.

The theoretical enemies of autonomy in ethics come from the "altruistic" ideologies propagated by some religions and conventional education. There is a conflict, not only in theory but in the field, in hospitals or parliaments, between the ethics of autonomy, usually modernist and progressive, and the ethics of "vulnerability," which can take different forms and denominations, but which in the end depends on a religious vision of the world. The sacralization of "the other" is a strategy that leads to submission; submission to instituted dogmas and concretely to the worldly institutions and persons administrating these dogmas. The key of this domination strategy is the dismissal of the self as debased: sinful and essentially "egoistic" as a consequence of the transcendent "fall" of the human. The self is then in a perpetual debt to "the others," perpetually guilty of being egoistic, which means in fact guilty of being a self. With this instilled guilt the self becomes indefinitely manipulable.

The "return of the self" in philosophy restores a positive and constructive understanding of the self. Autonomy no longer means solipsism or egoism. On the contrary, prejudicial views of a "sinful" self must be abandoned if we want to build up an active and constructive self, helping others and providing for their welfare and happiness, particularly through the wise practice of benevolence and the search for harmony. A resolute disruption with the Christian and Western obsession with "the others" is needed to clear the path of an active robust self, liberated from the domination and submission ethos which relies on moral shaming. In the crucial divide between autonomy and vulnerability in ethics, the choice of wisdom is autonomy. Vulnerability is addressed by the sage with practices of benevolence, plus humility, from the solid base of an autonomous self who never needs to be absorbed in serving others. A robust self does not search outside for its own consistence. Its behavior is rooted in the respect of the autonomy of others more than in the need to rescue their "vulnerability" and take them in charge.

Once rid of ambiguous domination tendencies and of any hidden religious agenda, medical ethics could be very different from the current paternalistic and formalistic doctrine. The picture is entirely different when it is centered on a self-reliant person who faces the questions of abortion or euthanasia, or of accepting a treatment, not in the ten-line account of a case in a textbook but rather in the long and original course of a human life. When the self is engaged in a self-construction process, or engages in this process at the occasion of a health crisis, its autonomy is the primal source of ethical legitimacy. Any other source would be pure and simple domination. If the self in question is apparently far from awareness and autonomy, far from being a morally self-sustaining person, the appropriate ethical approach is to help this person progress in the understanding of the factual and moral elements of her condition. An education toward reappropriating one's situation is sometimes required to confer the minimal autonomy that validates a person's decision. The method for this education exists in some abortion and euthanasia legal or deontological procedures. The important feature is that the turning point of the medical decision is situated in the educated self: it means that the self-construction process is esteemed to be more important than any external reference grid. When applying this principle, different stakeholders are involved; not only the direct subject of the medical condition. Medical practitioners, for instance, may not accept a course of action that they interpret as giving death, even indirectly. Relatives are implied and their moral education is sometimes a hard task. Future selves may be at stake, in the case of abortion or antenatal diagnosis for instance, or past selves, in the case of neurodegenerative diseases for instance: acting ethically in the face of these absent persons requires awareness and guidance. This is exactly why most cases of medical ethics are cases for Aristotelian *phronēsis*, applied wisdom in a very specific context, for which wise persons are the sole sources of solutions and solace. The ethical task is to constitute this sort of wise person, starting with oneself. In their absence, applying an abstract set of rules, from legal or religious origin, is a profoundly non-ethical procedure. Self-care and autonomy lead to the wise practice of moral self-defense.

In less dramatic instances, autonomy can be mentioned for payment card use. Taking care of one's payment device can be reduced to a simple question: does this payment give me more or less autonomy? With this awareness, small and big financial decisions correctly engage self-care and wisdom on the level of the ordinary. Similarly, smartphone apps for jogging, walking or swimming, monitoring fitness or sleep, etc. can be tools for autonomy and not gadgets. It all depends on the presence or absence of a consistent self and not on technology itself. Autonomy for energy (consumption and production) can be achieved with solar panels and windmills for home production, or any form of local production of energy, on the scale of a house or block of houses. It could be simple to solve the predicaments of fossil fuels, a major geopolitical nuisance, and the nuclear industry, a major environmental risk. Just by acting locally, we can avoid entering into global confrontations. Financing autonomously the solar panels and windmill, as some citizen groups do, disconnects us from the addictive networks of financial utilities – it prevents creating one's own "vulnerability."

The atmosphere is one of invisible confrontations and even moral wars. Two unseen antagonisms are entrenched in the technosphere and infosphere: one bears on time and the other on data.

The importance of one's time and of managing it in full awareness is another common precept of wisdom in almost every school. It is prominent in Seneca, whose civilization environment, the Roman Empire, was in a sense similar to our times (for the elite) as far as material and cultural abundance are concerned. In his letters to Lucilius, which are explicit lessons of wisdom, Seneca begins his instruction with the reappropriation of personal time (Seneca 64: Letter 1). Our time is stolen from us by all kinds of solicitations and distractions. For us, in a leisure society, anything distractive can be supplied in profusion, except the time to enjoy it. For this reason, the advertisement industry is not directly trying to reach our money but rather our time schedule. The problem for the film industry, for instance, is not to convince us that the last movie is worth the price of the theater ticket but that it is worth the three hours it will cost us to go and watch it. Similarly, the target for the tourism industry is not directly taking from us the value of a week or a month of salary but taking from us one week of our summer vacation. The ads on the walls of our cities and on TV screens are tracking second after second of our time and taking it by force.

Rifkin (1987) retraces the history and aggravation of "time wars." Protecting the time of one's life belongs to one's moral self-constitution. The practice of autonomy invites us to *make* time for important things that we do not *have* time to pursue – some of them being self-care, some being links with others or engagement with a community. Practically, autonomy means a capacity to resist excessive solicitation from work, from social activities, from digital addictions. Parkins and Craig (2006) depict the "slow" movements coming from Italy and from the resistance to "fast food." This culture war is entirely waged in the everyday and over ordinary issues such as food and transportation. The importance of the ordinary is clearly one of the philosophical traits of the movement.

The example of email overload and email management illustrates resistance methods. Many studies confirm that the majority of us spend far too much time reading and answering email, for professional or personal purposes – and for mixed purposes more and more frequently. A discipline and technique of the self, in the Foucauldian sense, can be put into action here. Its practical means are purely individual adaptations, such as using several mail addresses, to begin with. One of them is only for junk mail. This address is easily divulged when a subscription procedure intrusively requires an email address. A complementary measure is to use a "client" software (like Thunderbird) to retrieve messages on one's personal computer, instead of managing them on websites (technically: using POP and not IMAP protocol). Email software allows a parallel use of different email accounts, with different rules, and an easy definition of rules to filter the messages (not seeing what arrives on junk addresses for instance, or from spam or quasi-spam senders). These options are not technological only, they are ethical and wise, in the defense of autonomy. In the same line of argument, the autonomy of the self in terms of time and attention must be protected by an ad-suppressing add-on (complementary software) running with the Web browser (like AdBlock for instance).

The time spent in text chatting or online video conversation can be easily mastered by a wise management of the software options that make one "visible" or not on the network. Without any manifestation of real aggressiveness, the sage in the technosphere protects his or her autonomy by acquiring and exerting the technical competence to protect his or her time online. Saved and protected time, online and offline, can be a gift to others: it has a unique value in the technosphere and it is a much needed ingredient for benevolence.

Compared to the time war, the data war is more directly confrontational. There are identified enemies here: the corporations and the states that spy on us. However, I will argue that the defense of autonomy can still be non-confrontational self-defense.

The value of privacy needs a realistic reexamination of its content. In the most prominent cases, privacy is the form of autonomy that people must protect against state or commercial surveillance. What is at stake is actual domination: the command-and-control of human persons by institutions, through repressive or incentive means. This danger is amply documented in respect to smart ID cards, medical files, payment software, and in general the ubiquitous marketing and bank tracking possibilities which are already active in the infosphere. But what kind of autonomy exactly is it wise to protect?

Treating loss of "privacy" as if it were an intangible metaphysical asset is questionable. Disclosure of personal information *to whom and for what purpose*, and not a metaphysical right to privacy, must be examined. Contexts matter. Let us realize that when we pay with a card in a supermarket, the information system knows exactly what we have bought, in any revealing detail. But once we are aware of this, we can buy some items at a different place and pay in cash. We can reconsider what we really want to be private. Coevolution has it that privacy cannot stay within its pre-digital perimeter. Considering the big picture, the consequences of transparency may impose a more tolerant social environment, the tolerant environment that can precisely be built once people are aware of data transparency. Instead of keeping something secret, we could evolve the social environment into tolerating it, into being less judgmental, and the total move would be moral progress. I for one would endorse the following declaration: I confirm hereby that I authorize Amazon to know everything about my books and music purchasing behaviors – in exchange for information and an ordering and delivery service that I would have considered pure magic when I was an isolated youngster in a small town of southern France with no access to foreign books or classical music. Secret clause: I believe that I am smart enough to hide from Amazon what I want to hide and I keep an eye on technical evolution to keep it that way.

Sapiential technoethics is focused on the ordinary and then on *ordinary autonomy*: microactions that modify global contexts and in reality humbly command global coevolution. These microactions are wise variants in the use of the most ordinary devices: non-heroic autonomy. For instance, the same deep-freezer can be used for collecting and stocking raw materials (as far as possible in season and local) and for cooking from them, autonomously, or for piling up industrial junk food. A sophisticated home food processor or mixer can either remain an unused gadget or else become the focal and radiant source of homemade cooking, sharing, and healthy living. The Internet can be used to order pizza or to find recipes for homemade pizza. These microactions are not strictly determined by technical or social context, or more precisely: no other conditioning factor can supplant the resolution of a self-reliant self.

Ordinary digital autonomy is composed of analogous microactions and micro-decisions. For example, the "quantified self" practice consists in recording every data about oneself (biometrics but also location, consumption, connections, every activity) for the tied purposes of analyzing them and publishing them online. An autonomy-minded variant is to use simple tools (specific or homemade with a spreadsheet software) to manage what is important for oneself in a phase of one's life: let us say weight control. This can be a wise practice, neither obsessive nor addictive, far from gregarious "sharing" or absurd competition. The computer or smartphone can be of real help for improving one's weight and physical condition, through daily recording, planning, calorie counting, or inventive quantification of one's diet and lifestyle. The point is to appropriate the tool, whatever it is, in a practice of autonomy and wisdom.

Opting for backups of one's digital data on a local physical drive, and not "in the cloud" only, is a practice of autonomy. Software corporations try hard to sell online services, online storage, and online software. This option implies a total dependence on the service provider, in commercial terms as well as in terms of surveillance. Digital autonomy can easily be preserved on local machines and drives instead, to resist the insidious disempowerment which is now rampant in the infosphere.

Exactly as the capacity to read and write, minimal digital skills are necessary for autonomy in the technosphere. These skills are sapiential values. Rushkoff (2010) puts it all in the concise alternative: "Program or be programmed." It is a new duty to learn the basics in computer code, to master the basic notions about computer systems and programming. Simple codes like HTML suffice to understand what code is – or even the "macros" of current office software will do. Digital autonomy naturally demands using "open" software and document formats (such as LibreOffice and odt format), in order to join the fight for a free digital world but also, and nearer to the core of sapiential technoethics, in order to develop one's digital autonomy.

Wisdom implies waging the unseen moral wars of today, concerning autonomy, such as the time war, the data war, and any other micro-autonomy challenge. The practices of autonomy are self-justifying practices: with them the self is fighting to be a self. However, the barbarian language of "fight" and "war" must be abandoned: it was used here as an imperfect image only to convey the intensity of the required engagement. In proper terms, the practice of sapiential autonomy should opt for the Daoist spirit of "victory without fight"; it can only be a practice of non-confrontation. This style of action is possible in the present technosphere and infosphere, and, even better, it corresponds to their most promising potentials – flourishing responsive selves in a non-confrontational harmony of practices.

Harmony: Appreciative and Active Skills

Awareness and autonomy are already Western values, although they appear here in a reinterpreted meaning. On the contrary, harmony as a fundamental value and as a "practice" brings something radically new into the Western vision of the world and of human action. Harmony is strongly linked to awareness, humility, and benevolence. A *practice of harmony* would consist in two stages: (1) perceiving the immanent harmony in any situation; (2) taking action (or non-action) to enter into this harmony and possibly enhance it. This attitude is diametrically opposed to the Western technical pattern: (1) planning the wished state of affairs and mobilizing the maximum available means to impose it; (2) taking the most efficient and direct action to carry out the plan.

Harmony can never lead to gullibility, submission, unconditional acceptation, passivity, and all that Asian philosophies are reproached for. Rather than despair and resignation, harmony leads to engagement and transformative action. Martin Luther King and Gandhi were motivated by the representation of a civil and political harmony that did not exist at their time but could be realized by determined action. An accurate sense of harmony detects the absence of harmony and then engages in resistive action.

When autonomy is conceived as flat independence, it belongs to the Western command-and-control values system. Autonomy can also be integrated into *interdependence* and in this case it relies on harmony and authenticity. The sage, then, needs a deep understanding of interdependence: not only an abstract knowledge of systemic facts and laws but emotional perception of its entire symbolic and esthetic load. This acquisition is a practice. It belongs to the active practice of harmony in interdependence. Interdependence here refers to the network of human actions, particularly the emergence of global effects from microactions. It also means interdependence among every being in the ecosphere, from the most local environment to global issues like climate or non-renewable resources. Finally, this interdependence includes the third actor of coevolution, technology.

Chinese philosophy and Stoic philosophy abound in meditative views on harmony as a cosmic constituent (the *Dao* or the *logos*), infusing moral patterns everywhere. Ethical action then consists in tuning oneself with the cosmic harmony that founds the value and the beauty of action. The direct passage from cosmology to ethics and even to politics in these doctrines has been totally lost in modernity because of the strict divide between neutral objective reality and the isolated active moral subject. Contemporary technoethics does not call for transcendent cosmology, as in classical Chinese philosophy, but only for immanent practice and educated perceptive skills concerning harmony. In family life, in the countryside, in a pristine natural environment but also in an ecologically integrated urban or industrial project, harmony must preside over judgment, ethical or esthetic judgment, or, better, wisdom judgment. The identification of ethical value with "beauty" in ancient poetry and philosophy, the importance of beauty in the environmental views of Thoreau or Naess, and numerous cultural symbols indicate that achieving harmony relies on perceiving its beauty.

Surprisingly, Thoreau interpreted some technological features (the sound of the train and the railroad, telegraph wires) as genuine elements within the harmony of life in the woods. In *A week on the Concord and Merrimack rivers* he compares the sound of the wind in the telegraph wires to a divine melody, the "telegraph harp" (Thoreau [1849] 1985: 143). This is not a naive confusion of genre; it is a superior capacity to perceive harmony in the actual environment, in the continuum of nature and technosphere, without any prejudice. When Thoreau uses artifacts, be it a spade or a pen, the real concern is the harmony he is participating in. In Thoreau's utopia of transcribing natural harmony into literature, an ultimate skill is sought after: "A perfect work of man's art would also be wild or natural in a good sense" (Thoreau [1849] 1985: 258). From this perspective, the village road and the fields, artifacts that could commonly be taken as "natural," can legitimately be designated as so, or better they can be felt as harmony within the environment. They participate in the harmony that is the essence and value of nature.

Ancient Chinese philosophy considered entering into the harmony of the whole as the ultimate accomplishment of wisdom (Zhongyong 600 BCE: Section 23; Ames and Hall 2001; Angle 2009: Chapter 4). Appreciating cosmic harmony and actively joining it are essential practices of wisdom. In Confucian ethics, harmony is more than the desirable state of peace – the order of the Empire due to the universal submission of humans and nonhumans to their "roles." Harmony is not only a "rooted global philosophy" but also a "constructive engagement" (Angle 2009: 6). Beyond the conformist moral of harmonizing with "Heaven" or "Fate," today the stress must be on individual training for "harmonization-as-a-skill," a skill originating in ancient Chinese or Greek wisdom but newly invested in technoethics and rid of ancient cosmology. Wong (2011) states that the spirit of harmonization is perfectly applicable to the technosphere: the contemporary capacity for harmonization still consists in personal concrete skills for harmonizing a given complex situation. It gives a Confucian version of sapiential technoethics. Recent academic publications are numerous in this field (Angle 2009; Chan 2014; Hershock's books, as they draw much from pre-Buddhist Chinese philosophy).

Elaborating a "Buddhist response" to the infosphere's moral challenges, Hershock (1999) activates under different forms the same framework: interdependence and not independence. This orientation is presented as a Copernican revolution that addresses the real roots of suffering (the primary issue in Buddhism). In this context, Hershock puts forward a very original notion: *virtuosity*. Sapiential technoethics can draw a lot from the notion of *existential virtuosity*.

The perceptive shift that opens us to harmony, to begin with, is called "appreciative virtuosity" (Hershock 1999: Chapter 7). *Appreciative virtuosity* in the perception and realization of harmony is a powerful practice of wisdom, especially in a world of always changing and infinitely diverse realities, like the present technosphere and infosphere. Some of us may estimate that in this present world the average teenager shows more "appreciative virtuosity," that is more perception of possible harmony and harmonic development, in a new smartphone app, for instance, than the majority of scholars and journalists who tend to apprehend it from a technophobic background. In explaining virtuosity as the capacity to perceive interdependence, Hershock enlarges the picture (Hershock 2006: Chapter 8) and ultimately incorporates virtuosity into a comprehensive new form of practical philosophy (Hershock 2006: 191–192). This appreciative and inventive practice of harmony and harmonization is particularly welcome in sustainability issues. To promote responsible consumption within global interdependence, we need a culture of awareness with a strong valuation of the local and global harmonies, an active awareness of the ecological "cost" in the life cycle of goods and services. Eating less meat, possibly none, can be chosen because it is felt as a better insertion within the harmony of the ecosphere given the current state of the breeding industry.

Harmony makes sense in particular on the issue of energy, probably the most revealing issue in modern technology. There is a choice between harmony and violence in exploiting nature's energy. The idea of "soft energy" as opposed to "hard" is not poetry but a defendable approach to the difference between renewable and flow energy, on the one hand (the soft), and stock energy, which is more or less violently extracted, on the other hand (the hard). Some of Thoreau's essays about modernity ("Paradise (to be) regained" or "Life without principle") are structured by this distinction and they typically discard the disharmonic violence in modern industry's energy supply (Thoreau 2004: 19, 155).

Grinevald's account of our "combustible" industrial civilization sends a message about the violence of mining and burning, the violence in the irreversible appropriation and destruction of natural stocks (Grinevald 1976). From this perspective, a philosophical assessment of shale gas extraction can bring something new into the confrontation between nervous stakeholders: the consideration of harmony. Even if it is not worse than regular oil and gas mining, shale gas exploitation with "fracking" technology is an aggressive and violent disruption of the ground with irreversible consequences for its physical structure and chemical components. Therefore, such a technology is not wise. It may be an economic bounty (for some at least) but not a wise option. The end of the story depends on who cares for wisdom assessment and how it has a voice in the debate. Ironically, one of the best arguments in favor of shale gas exploitation is the harmonic coexistence with farming activities, some of them ancestral, in the USA (farmers lease their land and the complementary income helps them survive economically): even the fracking industry understands the value of harmony, it seems.

There is an "experiential gap" between energy consumption (in the proximal technosphere) and the complex chain of energy production and distribution (in distal and global infospheres). This "experiential gap" has been noted by Briggle and Mitcham (2009) and commented on by Geerts (2012). As long as the electrical network remains opaque, no "ethical" electricity consumption is possible. Closing the experiential gap creates the moral competence required to address the energy predicament at the appropriate micro-level on the user side. A moral experience of the global in the ordinary belongs to the harmony skills required by modern wisdom.

On the individual level, the harmony of the self with its own body and the harmony of this body with the environment constitute an important practice of wisdom. Being able to walk, run, cycle or swim, living with a balanced cycle of sleep and rest periods, enjoying the harmony of one's physical passive and active insertion in one's environment are all components of well-being. They are materially easier to reach in the technosphere than in any previous material environment. We are insufficiently attentive to these opportunities for harmony because our natural orientation is toward imposing our plan on reality and on others.

Humility: The Real Strength

The problem of arrogance is pervasive in Western civilization and in individual behaviors. An "innocent" kind of arrogance characterizes this civilization. It stems from ontological arrogance: taking for granted that the whole of nature, plus other civilizations, are at our disposal for the advancement of our own program. This method inevitably leads to confrontation and to missing the opportunities of harmony and coevolution.

On the level of the ordinary, a low form of arrogance is implanted in us: vanity. People seem to live as if they were constantly recovering from a previous humiliation. Actual humiliations may have been experienced, due to parental education, school, or the social system, all of them sometimes acting as if they intentionally want to humiliate people (Margalit 1996). We cannot easily change this pattern on the side of the perpetrators of humiliation because they are almost always unconscious perpetrators. The real change then must take place on the victim side: being aware of the mechanisms of humiliation and resisting them, internally and externally.

More broadly, this implies being aware of the logic of arrogance and vanity, and then refusing competition and confrontation in general. This ethical move is decisive for wisdom: it reorients a lot of behaviors in the technosphere and infosphere.

What social critique calls the "system" essentially counts on our vanity and it does everything to promote vanity. Let us complain only moderately in this matter because other domination systems use brute force instead. Resistance can be non-violent and even non-confrontational, but it requires a radical conversion to humility. In Table 1.3 the third and neglected resource, power over oneself, contains a particular form of humility. The humility of wisdom has nothing to do with masochism. On the contrary, it is an internal pride that delivers from the need for external testimonies of value. It thus delivers us from the domination pattern that is implanted in the cultural heritage of religious and moral humiliation of the self. Humility is not the love of humiliation but the indifference to it. It provides a natural resilience to aggression and manipulation through humiliation.

For ordinary technoethics, humility means indifference to the system of vanity that commands consumption in our civilization. Ordinary changes from vanity to humility are easy to conceive. The kind of car that people buy, compared with the average income of people in a given country, demonstrates that the car is a prestige (vanity) artifact before being a transportation device. Ordinary resistance is spreading in some countries, where the car cult seems to be receding – in France the low-cost brand Dacia, which was created for "poor" countries, met an unforeseen success without any commercials or prestige value.

An optimistic view of this change would be that humility could challenge vanity in terms of lifestyle and even lifestyle fashion. There is an interesting difference in the effects of these two satisfactions in a civilization of abundance: the satisfaction from vanity erodes very quickly (a new car will be needed soon, or a new professional promotion, or income growth), whereas the satisfaction from humility tends to be durable and even to increase in time. New sages take joy and pride in owning and using old inexpensive devices, perfectly maintained through time. They experience another sort of pride in these things, a pride accessible in a consumer society: the strength of wise consumption – as opposed to the weakness of compulsive consumption.

Mental exercises to achieve strength in situations of voluntary humility are worth trying. It is a rewarding practice and it can lead to unexpected joy – for instance in using public transportation, walking instead of calling a taxi, cycling instead of driving. A striking image of the tide reversal in technoethics can be imagined: a brand new SUV and a bicycle (one from a public rental system now frequent in European cities). In the last decade, the level of moral satisfaction and well-being has been reversed: it is high on the bicycle and not so high in the SUV. If a cyclist or pedestrian endeavors to "shame" the SUV driver, he/she would understand what is meant – a situation that would not have been understandable decades ago. The driver's arrogant interpretation (they are jealous of my car) becomes less and less probable, especially because the tidal wave of humility and frugality comes from the wealthier and more educated classes.

The same logic of simplification and voluntary humility applies to clothes: it is well known that the most admired new class of rich people, in high-tech business, wear jeans, T-shirts, and "hoodies," not expensive suits. It can apply to all sorts of personal belongings: "antique" mobile phones are fashionable in some circles. In some circumstances, to work on a computer older than five years will be seen as a mark of maturity and competence (indifference to commercial manipulation by the industry).

The logic of humility versus vanity can help solve some ethical questions bearing on the digital. The transition is natural because social status and prestige are more and more linked to what happens online. Social networking sites push the limits as they transform every individual into a self-promoting activist. Audience scores and the number of "friends" are often considered without much analysis as values in themselves. Every form of exposure or trash video seems to be an acceptable means for improving these scores. In Facebook's founding principles, in particular, reigns a typical American college immature mentality (I apologize for the simplification and stigmatization) that can be resisted on solid ethical grounds.

Thoreau's imperative "Simplify!" is then a universal resource for wisdom. The ethical practice of simplification, directly following from the humility option, allows the modern sage to resist the snobbery of complexity and to dare embrace simplicity. Simplicity is an ironic affordance of abundance, clearly underused and underestimated. It is possible to go to the mall and buy just exactly what one needs. The humble ordinary, in the sense valued by Borgmann about focal things (Borgmann 1984: Chapter 23), remains available. The motivations for it are simply unclear without the ethical awareness that I am trying to pinpoint.

Humility leads to a wise consumerism, a more reflective and responsible use of abundance. Abandoning the control meta-attitude is a basic precept of modern wisdom. Humility is destined to modify meta-attitudes in the technosphere. Behind this change of ethical style there is a subtle *liberty* to be discovered, which combines attachment and detachment. And this is Zen.

The ethical value of humility lies in the relationship between the self and every "object," material or immaterial. A *free* relationship must be conceived. The sage conquers it in dealing with the dilemma: attachment or detachment? When analyzing the positive and constructive potentials of technology we have examined attachment to artifacts, material or immaterial (Section 3.4). We have argued that sound attachment is possible for a consistent self. All the practices of wisdom described so far (awareness, autonomy, harmony) are compatible with a sound attachment to artifacts – including a connected telephone and a website like Wikipedia. A sound attachment to one's sources of food, comfort, and pleasure in life remains a value in technoethical wisdom, which shuns masochistic privation. But at the same time this basic existential fact must be combined with the recommendation for detachment that is found in Buddhism, Zen, Stoicism, and the quasi-totality of wisdom schools. Detachment releases the sage from suffering and self-destruction. The solution is to refuse the binary choice for a typically Zen superposition of attachment and detachment. The free relationship with a thing and ultimately with the technosphere is then: being attached as being detached – for instance to own and use a thing while being perfectly aware that it could as well no longer be owned or used in an instant. Other wisdom practices are active in this one. The *autonomy* of the sage is involved in this free relationship, checking that no addiction or simply habituation sets in. Then the appreciation of *harmony* makes the sage aware of the long-range and long-term consequences of certain behaviors or possessions; it can lead to moderate use, a form of detachment within attachment. For instance, one can appreciate hot showers without having four of them in a day or spending cubic metres of water.

The practice of humility as attachment-detachment is one of the most important skills for wisdom in the technosphere, requiring a capacity for improvisation and constant training. A paradigmatic existential virtuosity is needed to maintain a constant balance between self-construction through the affordances of the technosphere, on the one hand, and the practice of humility, on the other hand. The *Zhuang Zi* gives a political form of the paradox that helps understand it: "Only he who has no use for the empire is fit to be entrusted with it" (Zhuang Zi 1968: 309). This pattern is a paradigm: only the one who could perfectly flourish without a car or smartphone is fit to wisely own a car or smartphone.

This form of freedom and the practice of humility (in full awareness, autonomy, and harmony) constitute a somewhat addictive experience. It releases reward hormones at high doses; it gives a feeling of serenely dominating all possible situations, in total independence (within interdependence). This experience is diametrically opposed to the current experience of stress in modern life where preoccupation colonizes the entire mind and stress hormones submerge the body and brain. More precisely, humility is key to the experience of another sort of "control" (which is non-control, letting go, serenity), in a detachment allowing attachment. The satisfactions of attachment and those of detachment blend together very well, in fact.

Ordinary wisdom is far from heroic wisdom. It does not prescribe parading in the streets naked and covered in ashes (as Hindu sages literally do, totally without purpose according to the Buddha) but just ordinary humility – which sometimes can taste like parading in the streets naked and covered in ashes, I confess.

Benevolence: Realizing Non-duality

Benevolence is another neglected source of joy. A lot of religious and moral doctrines have professed this, but their applications have had varying degrees of success. Modern wisdom needs to rehabilitate benevolence as a fundamental virtue and practice to the point that benevolence becomes the number one value for inter-human relationships. On the one side, it will be limited by self-destructing naivety, and on the other side by paternalism – moral domination under the cover of "altruism." The latter concern requires a serious dewesternization of benevolence.

A renewed notion of benevolence can be drawn from Buddhist moral philosophy where the suppression of suffering in all beings is the first (and the sole) purpose. Every virtuous action is a benevolent action aimed at reducing suffering. Benevolence in this sense is identical to virtue and to wisdom – in Buddhism this perfect equivalence is the ideal of the *boddhisattva*. A more practical definition of this virtue mentions positively the "happiness" of beings, as in Harvey's definition of "lovingkindness": "the aspiration for the true happiness of any, and ultimately all, sentient beings, for all these are like oneself in liking happiness and disliking pain" (Harvey 2000: 104). But the heart of the matter is not a technique to induce joy in others.

On a more speculative level, benevolence follows from the fundamental doctrine of non-duality. Shantideva beautifully expressed the non-duality of suffering in this principle: suffering has no "owner," it is neither yours nor mine; it must be avoided and suppressed absolutely, not in its limitation to one person, be it me or someone else (Shantideva 763: 8/101–102). This foundation of universal benevolence merges with the practice of harmony in wisdom. When the self opens itself to the outside in a spirit of harmony and awareness of the whole, it meets the other selves. Benevolence then still applies beyond the range of the human person: it applies very well to nonhuman animals too. In a broader interpretation, it can apply to an ecosystem as well as to an infosystem. In all these cases, benevolence extends the self beyond the "narrow" (egoistic) self. The experience that the other's joy, pain, and flourishing are not "outside" the self is a founding experience for the practice of benevolence.

How can benevolence shape the relationship with others and with social communities in the context of the technosphere and infosphere? To explore this potential, I take a minimal definition of benevolence: predisposition to do good to others, primarily according to their own definition of the good, accompanied by a presupposition that these others are willing the same.

Real and thought experiments are simple and some of them are surprising. Let us imagine *driving with benevolence*. The experience is accessible to any driver at any time. It requires a lot of humility and a fine sense of harmony: for most drivers it would need a complete reeducation because their natural instinct is set on enforcing one's rights and privileges on the road. An absorbing mental exercise would be to see the other driver, or the pedestrian, as a self who is identical to one's own self, in its rights, emotions, imperfections and so on. A demanding self-controlled behavior would be to act according to a meta-attitude of doing good to the others. But the reward should be an intense feeling of flowing harmony, in the traffic itself basically, plus the absence of bad feelings, aggressiveness and stress, in one's self.

Doing one's share of the chores at home and in the workplace, provided that in the present technosphere these chores are made easy and healthy, can be turned into a self-rewarding activity. It demonstrates autonomy in managing one's environment and one's time. It gives to others exactly what is precious in the present world: care and time. As a practice of the self, domestic house cleaning or professional mundane tasks can be *samu* (meditation in ordinary activities) or even gym – because in the present world we need physical exercising and the opportunity offered by ordinary chores is very easily accessible.

The main possibilities for personal commitments to benevolence in the present technosphere rely on information and education, on human physical welfare, and on the protection and promotion of the commons, material and immaterial. These domains are strongly related and all of them are dependent on public action, which exceeds ethics. But as we have seen more than once, technoethics stands in for politics and public policy when they are defaulting. There is in the "welfare state" administration, in the health care and education systems, and in environmental policy a priority for collective investments and actions, but there is also a trend of freezing into dogmatic institutions working more and more for their own sake (see this logic of counter-productivity in Section 5.2). It happens that some teachers and some nurses are perceived as benevolent whereas the institution they work for is not. This is because microactions are taking over institutional action: it belongs to benevolence altogether.

The Internet can be seen as a reservoir of benevolence. This view is coherent with the history of the beginning of the Internet and with the remaining "open" and "generative" parts of it. The culture of sharing, legally or illegally, is deeply rooted in the Internet. Before any manifestation of *resistance*, the common manifestation of this culture is *collaboration*, help, service, from pure benevolence. Online forums, on any kind of subjects, demonstrate an astonishing benevolence, in the form of inventive pedagogy and limitless patience for beginners in the field. The usefulness of Wikipedia depends on the benevolence of a devoted community, partly, and on the benevolence of all readers, in theory, who are asked to improve any article when they detect a mistake or a weakness. An ethical rule is entailed in this process: one is personally responsible for any mistake one sees on Wikipedia because it was one's duty to correct it immediately. Witty minds who take pride in reporting Wikipedia's shortcomings have no idea they are reporting their own technoethical shortcomings – as far as the online practice of benevolence is concerned. This ethical rule simply restores the balance of benevolence: do not see the Internet solely as a reservoir (to draw upon) but as a contributive project (to contribute to, benevolently). Reporting a bug in an open source collaborative software is usually very easy online; it gives even the non-programmer a little share of the joy reigning in a benevolent community. There are malevolent people online too, of course, probably in the same proportion as offline. The "trolls" who spoil non-moderated online discussions can be seen, wisely, as a minority infuriated by the online realm of benevolence, for offline reasons.

The notion of *service* deserves examination from an ethical and sapiential point of view. To call it "service" gives an acceptable name to the interested benevolence of firms that supply us with medication, information, and devices of all sorts. In any Apple promotion video, the principal message is that this corporation is benevolent to people and not (only) prone to enrich its stockholders. Having *service* as one's principal means to make money is no ethical flaw and it is not contradictory with the main purpose of making money for the stockholder. Let us take Google as an example. In ordinary use, the benevolence of Google in terms of service is equivalent to Wikipedia's. The fact that Google is a for-profit company living on commercials and background sales of personal data does not interfere with the service. This perception is not absurd because, in the case of Google, a large part of the content depends on the benevolent production of ordinary users – publishing content online or linking to the best sites on a subject. Is this an inconvenient truth? An ecosystem of benevolent and for-profit services has nothing absurd or unsound in it.

I am worried by the fact that the term *service* is obsessional in business departments but hardly ever mentioned in human and social studies. Yet the technosphere abounds in services more than in anything else. We value services more and more consciously, including care and home help services. There has been a shift from products to services in the economy – what we want is the service of the car or the fridge, not to own the products. This analysis might excessively stretch the domain of benevolence but my point is to identify a blind spot in the system of values applicable to the technosphere: the value of service. Commodities and devices, independently of the marketing strategy to sell them to the masses, structure ordinary experience in the technosphere. We must pay special attention to *service providing* artifacts.

Wisdom in the technosphere, to sum it up, asks for a very broad and tolerant notion of benevolence. It sharply contrasts with the bad habit of anticipating malevolence in others. The alternative view of human nature, as collaborative and not competitive, that supports this stance on benevolence, considers the Internet as its proof of concept (Benkler 2011). But benevolence as a wisdom practice is not essentially a vision of human nature. It is an engaged activity that combines with all the other practices of wisdom, with the purpose of constituting a consistent self and of constituting, via interdependence and harmony, a better technosphere.

Courage: The Missing Resource

There can be no practical wisdom without courage. "Courage is preeminently the executive side of every virtue" says a classic in morals (Dewey and Tufts 1908: 411). Regrettably, in a world of abundance and facilitation we do not have many resources to foster courage. The situation is even worse because the sapiential courage needed for technoethics is not heroism, which rewards itself in social prestige or vanity. It lies instead in humble microactions, which remain unseen and unheard of.

Western wisdom philosophy, from the Stoics to the American Transcendentalists, celebrates courage as a paramount virtue, including the courage to be oneself. The courage to be a self is indeed essential for modern wisdom. A submissive and depersonalized possibility of living without ever being a self is offered by the technosphere. Once again, it is the ordinary form of courage that makes sense. Darling-Smith's essay on courage pinpoints, from her own philosophical background, the significance of this modern form or ordinary courage, with a focus on *daring*:

> The simple courage to dare means taking ordinary citizenship seriously, caring for neighbors, schools, social institutions, and the rest. Though the simple courage to dare need not move beyond the ordinary things shared with neighbors throughout the community, it involves risk because it is always possible to tell when you have failed in an invested friendship, family, job, or community role.
>
> (Darling-Smith 2002: 120)

In his tentative "domestic reform," Emerson discovered how difficult it is to change small ordinary things in one's life (Van Cromphout 1999: Chapter 5; Robinson 1993: Chapters 4, 7, 8). The courage mentioned here is moral courage, a specific form that often consists in social courage: resisting social shame and its power of humiliation, resisting social conformism.

Ordinary moral courage, a difficult virtue indeed, is necessary to change one's lifestyle according to one's reflection on values. It can be the courage to refuse or limit alcohol in some pressing contexts or other addictive technology for which social pressure is high. Not "being" on Facebook is hardly imaginable in some contexts. It can be the courage of daily body maintenance by exercising, naturally (walking or having a physical activity in one's profession or hobby) or with an exercise plan (like physical training in a given sport). It can be the courage to face a medical condition or simply the courage for preventive medicine and early screening for diseases (when early treatments are available). The courage to live with somber medical prognosis or heavy aftereffects is one of the new forms of courage for accompanying contemporary medicine's predictive and curative powers. Moral courage in sapiential technoethics is involved in accepting or refusing (both are courageous) some treatments or surgery that entail risk and physical or psychic side effects. All these forms of courage are cultivated by a self engaged in a rich and long-term self-construction process. A self entirely deprived of these moral skills will endure more suffering than necessary.

As it was implied from the beginning by the exclusion of heroic courage, ordinary courage is definitely non-confrontational. It has nothing in common with the energy that one can find in ideological enthusiasm and stubbornness. Even domestic reform cannot be a war against capitalism and the market, as it is for de-growth activists, because the ethos of combat very soon takes over the ethos of wisdom. Ordinary courage as a wisdom practice is strongly connected to all the other wisdom practices, protecting from ideology, and to a great number of wisdom complementary virtues and skills, protecting from stubbornness.

6.4 Existential Virtuosity for the Contemporary Self

Virtuosity

Every instance of wisdom in the previous Section may kindle this skeptic response: "It is not so simple!" I acknowledge this critique as an inevitable consequence of disrupting sterile scholasticism in ethics. I endorse it also as a paradox in wisdom, which is central to Zen schools: anything said about wisdom is wrong. In a modern theory of wisdom, we can take it to mean that anything said about wisdom is not a true statement but rather an indication for building one's personal path to a version of wisdom which will evolve with the self and which will be unique at any moment. I borrow Peter D. Hershock's term *virtuosity* and I expand its meaning to designate the reinvention of wisdom in its entirety. Wisdom is virtuosity as an incessant existential self-constructing activity. It is of course an acquired and not a given capacity; more exactly it is by definition a capacity that constitutes and perfects itself through practice. The best analogy is the virtuosity of a musical instrumentalist, a *virtuoso* (Hershock 2005: 40); the second best one is the martial arts master. Both have acquired a technique but transcend it: they possess it to a degree that frees them from the technique. When the virtuoso plays or the martial arts master is in action, *it is so "simple."* Wisdom is that kind of progression toward virtuosity. My response to the "not so simple" argument is in Zen spirit: the goal is to progressively understand (by practice and training) how simple it can be.

Virtuosity works as an intriguing notion in Hershock's philosophy. He is inventing the whole semantic for it in the process of coining the English for contemporary Buddhism. The Hindu and Buddhist notion of *samadhi* seems to be an important source. It designates the ultimate state of concentration and meditation, exceeding all analytical capacities of the mind. Hershock renders *samadhi* into "joyous and playful attentive virtuosity" (Hershock 2005: 2) and "attentive mastery" (Hershock 2012: 272). This typically Zen combination of meditation and action also inherits from the Ancient Western notion of excellence, *aretē* in Greek, which was the name for perfection in Ancient Greek ethics. The double dimension, meditative and active, is essential to the notion. It supplies technoethics with the subtle attachment-detachment capacity demanded by the present technosphere. The goal is to realize appreciative and contributory virtuosity (Hershock 2005: 69), corresponding to the *phronēsis* dimension of modern wisdom.

Existential virtuosity, then, combines two specific skills of modern wisdom: (1) the appreciation of differences, to let them enrich flourishing opportunities (Hershock 2012); (2) improvisation, the creative aspect of virtuosity, which leads virtuosity one step further than Western *phronēsis*. Creative virtuosity in wisdom means adapting and finely tuning action to the particular circumstances and harmonies of unique situations.

The Buddhist notion of *upaya* gives a precise idea of this practice. In Poceski's (2007) study on Chan Buddhism the original meaning of *upaya* is the name for the Buddha's pedagogical virtuosity in adapting the presentation of the doctrine (the Way, the *dharma*) to very different audiences, using an extremely open range of theoretical notions and symbolic examples. At first, *upaya* was a technique for efficient teaching, mentioned in the *Lotus Sutra* and in classics like Musō ([13–14th century] 1994: 21, 53–54). Pye (2003) comments on *upaya* as "skillful means" and in particular as provisional pedagogical means for the "unenlightened," a pillar of Mahayana Buddhism. Exactly as with virtuosity in practical wisdom and technoethics, there is a high risk with *upaya* of falling into overindulgent casuistry – the Japanese word for *upaya*, which is *hō ben*, currently means "expedient trick," recalls Pye (2003: 137). But in its highest interpretations, *upaya* gives access to the best of applied wisdom. Vimalakirti, for instance, was a lay householder and he practiced *upaya* in the sense of *samu*, meditative engagement with the most ordinary tasks of life (Pye 2003). For the sage, "the features of the world are all Buddhist teachings" wrote Musō ([13–14th century] 1994: 83). There is then no limit to *upaya*, either in the appreciative or in the creative practices of virtuosity. *Upaya* gives a model for the inventive adaptation of practical wisdom to the ordinary proximal technosphere: technoethical virtuosity.

Hershock (2012: 257–272) insists on the improvisational skills that are necessary to address modernity and on how Buddhist approaches can help. Pluralism and diversity are the resources for creative innovations in lifestyles and action policies.

> Embodying limitless *upaya* or responsive virtuosity is incompatible with adhering to any set principles or methods. Relating-freely in actualizing liberating interdependence among all beings is *not* reducible to making the right choices.
>
> (Hershock 2012: 289)

The interactive nature of existential virtuosity, in appreciative and creative performances, leads to an *existential conversation* with life. Life talks to us and we talk to it, in response and in anticipation, as in any conversation. Conversation skills must be acquired slowly and by practice. Exerting them implies a permanent awareness of the infinitely varied elements of the situation. Virtuosity in the art of conversation, perhaps an idle useless talent now, remains an interesting paradigm for practical wisdom.

Harmony must be active in the existential conversation. A conversation without harmony is a dispute, a quarrel. As the image goes, our present technological age is one of a dispute with reality rather than a conversation.

The notion of existential virtuosity bypasses the impossibility of directly expressing a "wisdom doctrine" in the form of "principles" ready for deductive application. Therefore, wisdom skills and virtues in this Section are introduced as invention and adaptation, improvisation, skillful means to progress in wisdom. No example or statement has a literal value; a different option can always be the wise one, by way of a minimal change in the context. Virtuosity bypasses the demand for norms.

Desirable Acquired Conditions: Serenity, Authenticity, Consistence

Human action is guided by the representation of goals. Ethical action in particular can be guided by the representation of desirable personal conditions. They are neither self-sustaining values nor duties. There is no duty to be wise, deontologically, and there is no guarantee that being wise brings about perfect happiness, teleologically. The desirable states of serenity, authenticity, and consistence can only have the secondary function of indicating our progress in wisdom and at the same time providing incentives for continuing. They can be taken as signs of normality in the process and the opposite conditions (anxiety, inauthenticity, inconsistence) may signal a problem.

When pondering, for instance, the question "To be or not to be on Facebook," in a precise moment of a given personal existence, the anticipation or the experience of serenity, authenticity, and consistence, or the lack thereof, could be decisive. "It makes me feel comfortable with my day when I have shared it on Facebook" is a sign of serenity. "I am feeling that the image of myself on Facebook is so partial that it is more and more a fake second self" is the sign of defaulting authenticity. "My answer to this post on Facebook is not in line with the person I really want to be" is the sign of defaulting consistence. Virtuosity is the dual capacity to appreciate these perceptions with the utmost clarity and to adjust personal existence with the utmost creativity.

Serenity is composed of harmony, autonomy and awareness. Even when the circumstances of life make the good life and happiness utterly impossible, the existential virtuosity to maintain serenity remains – in perceiving the still possible harmonies in autonomy and awareness (enabling detachment).

Heidegger's conference about serenity (*Gelassenheit*) has much to say on the present technosphere in ethical terms (Heidegger 1959). The German word *Gelassenheit* means "calmness" in the sense of release, literally "letting-go," which could be the New Age translation – Heidegger actually says in the conference *"losslassen"* (Heidegger 1959: 22), literally "letting-go." Interesting suggestions also come from the familiar use of "cool" and from the objectionable but familiar use of the adjective "Zen." The meaning of "easy" in Californian English designates correctly easiness and facilitation as the major characteristics of the present technosphere. However, Heidegger's approach is not exactly Californian. His lesson on serenity is at first sight rather abstract but in the end it leads to the technoethical idea that a serene relationship is a question of *distance.* Let us envisage the right distance with the artifact, a distance that generates neither anxiety by pressure (too close) nor anxiety by absence (too far). *Right distance* can be applied to the relationship between human beings and it can even provide a definition for respect. It applies particularly well to ordinary human relations, as in a family or in the workplace. Heidegger does not mention such empirical developments. Instead, he insists on meditative thoughtfulness and the feeling of gratitude in *Gelassenheit.* His well-known critique of the confusion between thinking and computing then launches a technophobic account of modern science and technology, *Gelassenheit* being for Heidegger what should be restored in opposition to modernity.

In a technophilic revision of this analysis, *serenity* will be the project to find solace, *Gelassenheit,* in the midst of modernity through a wise relationship to the technosphere, relying in turn on a wise relationship to oneself. Serenity is accessible to responsive persons who regulate and tune their distance with the artifacts of the technosphere. There are strikingly Zen elements in Heidegger's vision of serenity in the technosphere (Heidegger 1959: 23). When defining the serene relationship to things ("die Gelassenheit zu den Dingen"), he calls it simple and peaceful ("einfach und ruhig") and says it contains at the same time a "yes" and a "no" to the technosphere ("Haltung des gleichseitigen Ja und Nein zur technischen Welt"). Heidegger points out the importance of a meditative understanding of the simple in the process of discovering serenity: "For us humans the path to the near is always the remotest and then the most difficult. This path is the path of meditation" (Heidegger 1959: 21–22). The practice of *samu*, the meditative reappropriation of the ordinary, makes perfect sense in a dedramatized interpretation of Heidegger's *Gelassenheit.*

The Confucian classic on "balanced practice" (*Zhongyong*, sixth century BCE) pictures wisdom as a form of serenity coming from the capacity to keep everything "easy." It comes from an idea of the world as a process of immanent transformation, ignoring the transcendent and heroic power of a subject imposing its plans on reality. The active capacity for letting-go would then resist the obsession of command-and-control in modernity. This letting-go is a meta-attitude, as we have seen with the other practices of wisdom. It is more active and efficient than the obsession with control, which constantly fills the mind with preoccupation and leaves no place for availability and creativity. Because of this mental unavailability, preoccupation leads in reality to passive acceptance and submission whereas serenity can be active and it gives a robust background for personal assessment and possibly for resistance. Clearing out one's mind and maintaining serenity is a precept for the martial arts masters in the midst of action.

Beginners' exercises in serenity can start with the simple experience of slowing down, a version of *samu*. In any activity, just doing it very slowly brings an immediate and often spectacular gain in awareness, mindfulness, appreciation of life, and joy. This ordinary serenity provides a different experience of time and activity, a different experience of the self and of the self's capacity for reappropriating any life experience, living it with intensity. Any activity can be the occasion of this slowing down meditative practice: walking out to buy something, preparing one's breakfast, eating, putting one's clothes on, writing an email. Just doing it very slowly transforms the nature of the activity. It turns the moment into serenity through the peaceful joy of the reappropriation of one's existence. When motivation is low, when an activity is due but we do not feel like acting, turning it into a meditative experience of radical slowing down may help give a new motivation.

Pirsig's (1974) book, linking Zen and motorcycle maintenance, has become a cult book for open-minded techies because it gives this existential virtuosity a deeply original and personal expression. It shows that human existences in turmoil and distress, from an external point of view, can find solace in a Buddhist experience of modernity and of technology in particular (the famous motorcycle).

Authenticity is a self-founding value. The universal reason to take wisdom as one's objective, for a self, is that it is the most authentic way to be a self. Simply stated: the wise self is more self than the others. Since we have situated the decisive field of technoethics in ordinary life and practice, the Heideggerian and Borgmannian critiques of inauthenticity in the technosphere can be challenged if technoethics reveals the possible authenticity in the ordinary of modern life. The turning point is in the self and its capacity for resolution. In Stoic philosophy of wisdom and in martial arts, the importance of authenticity in the smallest things is paramount. Exerting this authenticity in the ordinary is an essential practice of moral self-care. Reciprocally, only a trained moral self is solid enough to care for authenticity in the smallest things.

The Chinese classic on "balanced practice" also pointed out *sincerity* as the key to perfect virtue. "Sincerity is that whereby self-completion is effected, and its way is that by which man must direct himself" (Zhongyong 6th century BCE: Section 26). This virtue is perfect because it concerns the whole and thus includes the practices of benevolence and harmony: "The possessor of sincerity does not merely accomplish the self-completion of himself. With this quality he completes other men and things also" (Zhongyong 6th century BCE: Section 26).

The fake logic behind the "star system" in the entertainment industry and its small-scale reproduction on Facebook prove the ordinary neglect for authenticity in favor of "glamor" or its petty imitations. This logic is not without alternative.

A technoethical approach that calls for a strong notion of authenticity is included in the transactional model of change, which has been outlined above when discussing coevolution (Section 2.4). Significant changes and micro-changes can be seen as transactions, with gains and losses, real and symbolic. Authenticity requires that these transactions are accepted in full consciousness and full moral conscience. Balancing gains and losses is typically a question of *phronēsis* (Aristotelian practical judgment), that is to say it belongs to technoethics as wisdom practices. Borgmann suggests an original conception of this balance in moral transactions inside the technosphere:

> Although this will invite charges of Luddism, dystopianism, romanticism, and worse, it needs to be said that there are *always* moral losses in any kind of moral commodification. But in constructive applications, the moral gains *unquestionably* outweigh the losses.
>
> (Borgmann 2006: 156)

The self builds authenticity in the flow of transactions that constitutes a permanent conversation between Technosapiens, nature, and the technosphere. Wisdom, in authenticity practice, simply demands that this conversation be sincere.

Psychotropic medication exemplifies this issue. For reasons that can be physiological or due to life circumstances, the modern self can be faced with the decision to take psychotropic medication (any psychoactive substance destined to modify mood, behavior, or mind capacities). The change here concerns the self itself, which is confronted with an issue of authenticity: can it be preserved and even restored, with the medication, or will it be lost in medication? The nature of Technosapiens favors the conclusion that I am more myself with my clothes and my glasses than without. In a lot of circumstances, I am more myself with my computer than without. This does not impose the conclusion that any psychotropic medication in any circumstance will promote authenticity but it sets the stage for *phronēsis* or existential virtuosity: in the infinitely particular circumstances of this precise moment in life (for instance, the postpartum period, a period of treatment for cancer, or a degenerative illness), and given the known or probable effects of the treatment, is psychotropic medication an acceptable transaction for the self? The answer is provisional, leading to a testing period that allows an empirical moral assessment of the treatment. No abstract rule can determine when and how a psychotropic medication can be inserted in the constitution process of a self, preserving or restoring its authenticity, or on the contrary ruining it. Sapiential technoethics suggests the opposite approach: a self engaged in a permanent examination of its authenticity, a self already trained in the existential virtuosity of authenticity, will be better armed to face this choice.

Consistence has been largely defined and examined above in the perimeter of the self (Section 4.2). As a wisdom virtuosity, consistence appears to be the result of precise practices: *courage* because the courage to be and remain oneself is sometimes extraordinarily demanding in ordinary instances; *autonomy* because the inconsistent self is easily torn to pieces by the never-ending solicitations of the technosphere; *harmony* because consistence ultimately depends on both internal and external harmony and the capacity to maintain it.

The technosphere challenges the consistence of the self, on the level of the ordinary, through facilitation and temptations of all sorts. For this reason, ordinary technoethics is focused on the robust self's capacity for resistance. Consistence is also the capacity to "walk the talk," that is to say to implement one's values in real life. The consistent self does not split into an arrogant moral self who does the talking and a pathetic acting self who always finds a reason for not acting.

Technoethics' goal is to promote the consistence of a self who is not fully wise but *proficiens*, progressing toward wisdom on its own way and within its own specific circumstances. Danah Boyd's constructive vision of teenagers on social networking sites is an instance of existential virtuosity (Boyd 2014). Young people are compelled to acquire specific skills for the smartphone: these are not technical skills only but also existential skills concerning the different communication channels and the complex social consequences of one's behavior on these channels. These skills probably remain impossible to detect and to understand by parents, teachers, and apparently by some social researchers. Many adolescents are nevertheless driven to maturity by the dilemmas, the trial and error experiences, the ambiguities and also the magnificent accomplishments that are lived through screens and smartphones.

The consistent practice of "no logo" is a revealing ordinary case. It is amazing to watch intellectuals, who pose as irreducible opponents to corporations and marketed-technology, read their text from an Apple resplendent McIntosh, with its apple logo flashing on the back of the screen, without realizing, apparently, that they nullify (in the ordinary) their grandiose social critique – at least for those who consider Apple's "proprietary" policy as typical of corporate nuisance in the technosphere. A minimal "no logo" ordinary action would be to stick a black sticky tape masking the brand and logo on a laptop computer or mobile phone. It takes one minute and costs nothing. It personalizes the device with a discreet ethical ambition. There is a message in it, meaning more than yet another brilliant critique of corporations' marketing, perhaps. Systematically opting for "no logo" (no visible brand) clothes, shoes, glasses, personal gear, cars, everything that is proudly "branded," remains open to all. In this practice, the self proves it is consistent enough independently of commercial brands (not of their products): this self does not need brands and logos to be consistent. This is symbolic reappropriation of ordinary technology and a form of ordinary inclusiveness because brands are, as they say, "exclusive."

Desirable Acquired Capacities: Integrity (Honesty), Generosity, Resistance

Integrity, mainly in the sense of *honesty*, is directly connected to *consistence*, examined above. "Integral" and "integer" stem from the same root meaning "whole." The added dimension is *benevolence*, a virtue that is not always prominent in theories of the strong self, but which must belong to modern wisdom, notably as a dewesternization feature inspired by Buddhism.

The divide between "real" life and life online affects consistence and integrity. The usual approach is centered on the consistence of the "two selves": you would never steal a CD in a physical store, then why do you feel morally authorized to illegally download the same music online? This question presupposes a strict equivalence of "property" and "theft" off-line and online. On this basis only does the question of splitting the moral self emerge. But as soon as this approach, defended by the media industry only, is found to miss the point, there appears for the self a more realistic question about integrity in the sense of *honesty*. In a majority of computers, there are illegal files, music or video files downloaded from the Web, or illegal copies of software – games and a lot of professional software. This permissive digital moral does not attest to a natural immorality that can only be repressed. Off-line or online, do people systematically steal when they know they will not be punished? Reality and moral situations are more complex. People still buy recorded music and movies, materially or online, they still buy books, even when the content is available for free, illegally, online. In the most interesting case, people test a new musical CD, or movie, or TV series, through an illegal download channel, then they buy it – not all the time, by far, but when they want to materialize their gratitude for the work, when they want to "own" it or to offer it to others, to share it as a present. Books that are made available for free online usually sell more in physical form (when available). Field studies normally demonstrate that downloaders are also buyers in majority: there is only one population. This population has a specific practice of integrity and honesty, linked to autonomy and even to resistance. For there is an acceptable argument holding that after having been prisoner of the media industry, prisoner of its choices about content and arbitrary prices, the public is now in a position to determine, in full knowledge of the cause, the contents it wants to access to and to pay for. Instead of a feeling of dishonesty, the downloader-and-buyer has a feeling of escaping the dishonesty of the media industry and of restoring a more balanced and authentic relationship to the artists and producers of content (*pace* the few billionaires among them who are industrial stakeholders and not only artists).

Evidence for this new perception of integrity is massive in online peer-to-peer economy, on websites like eBay. In the absence of legal enforcement, people deliver the goods that they sell online, and they even accept reimbursement and exchanges, sometimes more easily than commercial firms would do. The booming success of these sites would never have been possible without very high trust rates. The system of "reputation" of the sellers (assessed by the buyers) buttresses this new form of integrity. The moral atmosphere of these exchanges between "peers" is a remarkable form of non-institutional integrity. It relies on understanding that peer exchanges rest on mutual trust, which means understanding harmony in a benevolent network. Some of the most important practices of sapiential technoethics are already at work in this phenomenon. Instead of focusing on the rare and pathological cases of personality splitting disorder that are due to a pathological use of social networking sites, it makes sense to focus on *the ordinary integrity* that sustains a whole economy of online commercial exchanges between peers. The modern sage finds a moral reassurance in this humble integrity – AirBnB and the booming "Uber"-economy give even more massive evidence for this.

At a totally different level, integrity is required in a technological world for engineering ethics. The paradigmatic case (in the USA), the "whistle-blower," is not central for sapiential technoethics because whistle-blowing concerns extraordinary cases, by definition, and not ordinary uses. From space shuttle explosions to nuclear accidents, the literature in engineering ethics is based on the extraordinary. Reading this literature, one may have a feeling that something is tragically missing: a self, a consistent self, the integrity of a human person, with its associated virtues, courage (to react and to act), the sense of harmony (seeing the whole), benevolence (beyond one's direct interest). In safety-critical technologies, such as aircraft and air traffic control, mass transportation, nuclear, missile, and medical technologies (Tavani 2004: 105), ethical codes are in place but what is their weight compared with the presence or absence of a consistent self at the determining moment and place?

The ethical virtuosity of integrity in the technosphere can be active on the level of the ordinary even in engineering ethics. The problem is to ethically engage the ordinary, including at the workplace. Ordinarily, technological and ethical vigilance can be low. As long as the situation is perceived as "ordinary," it may (wrongly) be taken as "normal," in the sense of being deontologically or ethically acceptable. This is obviously an ethical flaw. To hear one's neighbor beating his or her partner or children can be "ordinary" without being acceptable at all. In the major ethical scandals that occurred in various industries, in oil extraction, finance, and transport for instance, the decisive actors were acting "ordinarily" in the sense of normally for their profession, following the common practice without analyzing it in depth. This particular form of ethical submission gives birth to the "banality" of evil in technoethics as in any other context. Evil is ordinary, it remains unseen in the ordinary, and resistance must tackle it in the ordinary.

Generosity is the capacity for proactive benevolence. It is not limited to "giving" in the sense of charity, although this sense can be important in some contexts. Another ethical dimension of generosity deserves to be recognized and valued: the virtuosity concerning *diversity*, beyond the simple tolerance of variety in human cultures. An ethical progression from the traditional virtue of tolerance to virtuosity as the art of including diversity is called for. Hershock's (2012) distinction between variety and diversity can be illustrated in the technosphere by the cohabitation in the center of our cities of car owners and non-owners (people who use public transportation, cycle, and walk). The minimum requirement of decent technoethics is reciprocal tolerance: accepting variety. But diversity goes further, in the direction of an active ecosystem in urban transportation, exploiting the potentials for harmony and inclusion that are latent in variety. The infosphere already offers car sharing sites and apps.

Technological generosity can be found in answering emails even when there is nothing to expect from doing it, in collaborating in forums related to one's domain of expertise, in publishing online material of general interest – not for vanity reasons, so possibly under a pseudonym or anonymously. In every milieu (family and friends, business, academy) leaving an email unanswered is a mark of defaulting generosity. The meaning of non-answering is that not a minute (or 20 seconds) of the recipient's time is available for the sender. Here the technological facilitation of communication redefines the moral context. It was acceptable not to have the time to write a letter and post it with a stamp. It is not so acceptable not to have the ten seconds to browse an email and the ten seconds to answer it – if only by a standard pre-written message that is sent in a click.

Resistance follows directly from courage and autonomy, composing a profoundly original capacity that characterizes modern wisdom. Political resistance as we know it remains the inspiration for the micro-resistance in a sapiential technoethics. However, resistance is a special case because it cannot be construed as a value in itself. To resist can be psychologically self-rewarding and even socially praised but these appreciations do not found an acceptable value. Resistance cannot be a lifestyle and an autonomous vocation, as it appears to be in some postures. From a sapiential point of view, resistance has meaning and value only when it is rightly motivated. In the general case, it can be rightly and strongly motivated by the unacceptability of what is resisted.

"Ethical resistivity" in the self would be analogous to the physical property of (electrical) resistivity: an obstacle to the circulation of what passes through a person. Not electrical flows but rather moral, political, economic, and ecological flows pass through human persons. A remarkable part of "ordinary evil" passes through human persons without their having a real conscience of their "passive agency." In Foucault's last philosophy, an "ethics of the self" may be the most urgent political task because the relation of the self to the self is the pivotal point of *political* resistance (Foucault 2001: 241, my translation).

Since it is justified only when facing the unacceptable (outrage and indignity), resistance belongs to the system of values of *acceptability* and not to the system of values of *justice* that is the reference in the West. Dignity (Margalit 1996) and acceptability are weaker notions than justice in a meta-ethical sense: they apply to lesser cases and they do not bear such heavy consequences, theoretical and practical. Yet they could be stronger notions in a sapiential sense. Fighting injustice sounds more and more ideological and it always risks ending in ideological confrontation: people fighting for justice from their own point of view. Fighting indignity is quite different. In practice, resisting micro-indignity is an important capacity for wisdom and it can be non-confrontational.

Resistance in the technosphere thus becomes an *ethical self-defense*, which can genuinely refer to the modern martial arts paradigm as a self-defense ethos. Its basis is the capacity to fight efficiently but the key notion is the ethical virtuosity to detect when fighting is legitimate. In practice, this ethos of self-defense is almost always the capacity not to fight, as a result of reflection, self-control, and training.

Charles Ess discusses in several books and articles the question of resisting the cultural globalization imposed upon every culture by electronic media. He finally falls in with a "soft determinism" (attributed to Don Ihde):

> This middle ground recognizes that although technologies certainly embed values and bias users in specific directions, these are not irresistible forces. In the jargon of Star Trek, resistance is not futile.
>
> (Ess in Hershock et al. 2003: 518)

The slogan "Resistance is futile" comes from the "Borg" in the Star Trek series. They are cyborgs (mix of humanoid bodies and electronic technology), obsessed with "assimilating" all the other species (their second favorite slogan is "You will be assimilated!"). The typical feature of the Borg is their total depersonalization: they are brainwashed and they act under the total telepathic control of a central node. Resistance against the Borg, a clear metaphor in the technology debate, can be seen at first sight as resistance against dehumanizing technology – which already provides a coherent and philosophically rich interpretation for a TV series. A second interpretation, centered on the loss of the self, is nevertheless interesting. Actually it is the only moral issue. In this second interpretation, the self-reliance and moral heroism of the main characters of the series (those resisting the Borg) are not just forms of Hollywood heroism: their attitude is typical moral resistance in the defense of the self as an unconditional value. The Return of the Self again.

In a philosophical article referring to the same slogan, Barry (2006) claims that "Resistance is fertile," from an economic and ecological point of view this time. He reaches a conclusion on consumption as resistance: "Indeed, one of the most powerful and radical political acts an individual or group can do in modern consumption-oriented societies is to refuse to consume" (Barry 2006: 39).

Microactions thus emerge as the privileged means for economic and ecological resistance in the technosphere, and particularly micro-non-actions. A subtler attitude can be imagined. Verbeek's (2013) article claiming that "Resistance is futile" does not oppose this latter view, but it goes deeper into it. It suggests an original philosophy of *evaluative non-confrontation*, I think. Verbeek introduces a version of resistance as existential virtuosity in the form of "accompanying" modern technology. This accompaniment is a critical stance but it differs from the age-old "dialectical" resistance to political oppression.

Aaron Swartz, a recent and vivid instance of resistance in the technosphere, exemplifies my technoethical grid. Swartz was the hero of a hacking epic that was harshly fought by the US judiciary system, leading to Swartz's suicide in 2013 (Knappenberger 2014). By meditating on the life and actions of Aaron Swartz, one can draw a line between two methods of resistance. On the one hand, Swartz was justified in his indignation (essentially about copyright for academic research papers, eventually about surveillance as well), and he was brilliant for broadcasting this critique. On the other hand, he may have been tragically short of wisdom both in public action and in personal life. Sapiential technoethics wants no other tragic stories like Aaron Swartz's, even if his accomplishment to block objectionable laws was useful and heroic. Extraordinary but tragic destinies are not necessary if the millions of anonymous users progressively engage in an *ordinary resistance* spirit.

There is a heroic and extraordinary dimension in the classics of resistance but this may not be the essential dimension of the message. Thoreau's civil disobedience (Thoreau [1849] 2004: 63) is based on the experience of one single night in the Concord jail in July 1846 in a friendly atmosphere: a case of "story-telling" in the contemporary sense of the term. The two years in the Walden woods cabin (right next to town) were also a symbolic episode, the substance for literary narration. Thoreau's authentic message is disseminated through all the microactions of his life, in their humble but consistent accomplishments and in his literary act of celebrating and transmitting them (a practice of benevolence and generosity). Nothing extraordinary in this resistance.

In Thoreau's lineage, the Gandhian spirit of resistance is disseminated among the millions who rallied, marched, and refused to buy British items or to collaborate with British administration. It is an *ordinary resistance* spirit, put into action by anonymous individuals. In modern times, Arne Naess theorized and practiced this form of Gandhian resistance (Naess 1974). In the same vein, the fall of the Berlin Wall in 1989 was the final moment of ordinary resistance by East Germans, in particular those who "voted with their feet," anonymously, and all those who anonymously resisted the regime through microactions of non-collaboration. It may be excessive to affirm that Microsoft or (better) Halliburton has the same kind of power and support than the former East German authorities, or than the former British rule over India, but even if it were the case, they could be victoriously resisted by microactions.

Chinese classics repeatedly advise the sage to refuse any collaboration with an unjust regime and to even refrain from visiting or staying in the country. The modern sage would not visit countries where a minimum of justice and democracy does not reign. It may be a form of compassion for the victims (the oppression of women, or the quasi-enslavement of migrant workers, or the violent occupation of invaded territories), or a form of refusal to fund unacceptable regimes, or, more importantly, a form of moral condemnation expressed in a non-confrontational way. Humility and courage are required for choosing another destination for holidays and for declining professional or personal invitations. Another reason may be given at first, in order to remain non-confrontational. But consistence requires one to be explicit about one's resistive moral decision when one is directly questioned about it. The tacit moral acceptance of abominable regimes by the flows of tourists and business people in some countries is a lost chance for ethical resistance.

Numerous e-practices already mentioned are ordinary acts of resistance to the dereliction of the Internet. They must become second nature: resisting spam and unsolicited advertisements, protecting one's privacy (encryption of personal documents and secured erasing).

Hacking deserves special attention. A very general form of "ethical hacking" can be considered a wise practice in ordinary resistance. For instance, when it is compulsory to disclose personal information, such as an email or a physical address, in order to access a service that should be freely accessible (it can be wifi in a hotel room or in a public place) or for accessing a service on a website, the modern sage would systematically give a random chain of characters instead of the required information. It works most of the time and when it does not a "junk" address will do (an email address never retrieved nor used). The same rule would apply to the leonine contracts or conditions of use (often not even "human readable") that one must accept with a simple click to proceed on some sites. When a real service is provided, in accordance with one's best-informed judgment, one may accept to disclose information about one's age or place of birth if the service provider values that kind of information: an acceptable transaction occurs. When dozens of pages of abusive conditions are imposed in a take-it-or-leave-it mode, resistance is morally justified.

As we have seen above (Section 5.3), hacking, in the general sense of a technological bypassing action, can force inclusion. A general ethics of hacking extends this logic to any relationship with institutions. It consists in the non-violent voluntary diversion of systems and institutions, by detecting and using what can be diverted for the purposes of an independent project, possibly contrary to the agenda of the system or institution. The history of the Web, in its origin and in the development of its open domains, follows from this "soft hacking" principle. The most inventive uses of technology, from SMS messages to online dating platforms, emerged from marginal generative uses.

*Necessary Sensitivities: Compassion, Empathy, Aversion to
Violence*

Modern wisdom relies on specific capacities to perceive values and non-
values in the environment. The dereliction of the modern self is largely a
loss of ethical sensitivity. Inspired by virtue ethics (including its "sentimen-
talist" versions) and by a Buddhist-like dewesternization, the modern
wisdom that I try to compose draws much from *compassion, empathy*, and
the *aversion to violence*. These sensitivities contribute to restoring emotion
as a legitimate dimension in wisdom practice. The sage tries to reach an
existential virtuosity based on ethical sensibilities. The metaphor of playing
an instrument, and trying to reach virtuosity, applies particularly well to
the self's exercising its ethical emotions.

Compassion is sensitivity to the suffering of other living beings. It originates in the wisdom practices of awareness and benevolence, combined
with sensitivity for harmony. Its application in technoethics will also
require the wisdom practices of humility and courage.

It is well known that compassion is the starting point of the Buddha's
awakening and teaching: nothing is more important than to free all sentient creatures from suffering (Rahula 1959; Harvey 2000: Chapter 3).
This includes the self. The essential link with harmony comes from
the doctrine of non-duality: suffering is not attached to a particular existence (mine or any other); suffering has no "owner" (Shantideva 763:
8/101–102). Compassion is simply awakening. It is a core duty in Buddhism and not a supererogatory praxis (Ives 1992: 33).

In ordinary technoethics, and with special reference to the practice of
samu (meditative ordinary activity), instances of compassion abound in
sustainability issues, particularly in energy and lifestyle issues (such as
cooking, eating, drinking, mobility, domestic comfort). These ordinary
existences in the technosphere can be practiced as exercises in compassion.
Ecological, economic, and political compassion motivate a reform of self-
behavior tending to better sustainability. The moral atomization of the
technosphere becomes unacceptable: we cannot ignore that our lifestyle
adversely impacts the social and personal situation of billions. The flows of
matter, energy, and money in the technosphere are ethically laden and they
pass through us, undetected if there is no compassion to feel them. Compassion is thus a requisite for the resistivity evoked above. Compassion
provides the emotional and ethical links between ordinary lifestyle and its
distant consequences. Consuming less fossil fuels and eating less meat can
be motivated by simple compassion. Changing the structure and values of
local communities and larger political entities can be a duty of compassion.
Since compassion is a strong emotion and since it can be induced and
driven by voluntary reflection, it can be an efficient practice of ordinary
wisdom, improving the technosphere from the bottom.

Empathy could be taken as the more general emotion out of which compassion is only one form. In its long history, starting with British moral

empiricism (David Hume and Adam Smith), *empathy* tends to stand for the moral sense in itself. Moral sensitivity in the technosphere, in a context of massive communication and benevolence, can actually reside in a thin but universal layer of empathy extending all over the infosphere – I have mentioned Rifkin's vision of an "empathic civilization" (Rifkin 2009). Empathy is focused on positive emotions: sharing joy and pleasure, social and intimate bonds, all of them practices that can be cultivated up to virtuosity by the modern sage. Easy communication and material abundance speak well for a culture of empathy in human flourishing. It does not harm the culture of autonomy, so important for the self.

For all these reasons, far from the technophobic fear of electronic media producing isolation, it is wise to develop links of digital empathy. An empathic global community fosters the experiences of sharing joy and constructive optimistic projects that are resources for self-reliance and serenity. The "negative" empathy created by tragic events (natural disaster or terrorist attacks for instance) demonstrates the positive empathy that is possible through electronic media. Certainly we currently underestimate and under-use it. In some families or communities of friends, a high capacity for emotional sharing has been developed through electronic media – video conversation in particular. This form of empathy is specific to the technosphere and infosphere. The modern sage does not question its authenticity but tries to reach virtuosity in its capacity to share authentic and positive emotions in the infosphere.

Aversion to violence is not directly non-violence but it is the necessary sensitivity to be prepared for non-violence. A central unseen issue of the current technosphere and infosphere is the unavowed cult of violence. Representations of violence in fiction give the key instance, in the infosphere, of what must be resisted by wisdom. Murder, torture, rape and equivalent extreme violence are seen every day on TV, in the news and in fiction. It is possible to educate one's moral sensibility into tolerating this representation of violence. My point is that the opposite sensibility education, a radical aversion to violence, including an aversion to its ambiguous representation, is also possible, advisable, and wise.

The "pitch" of fictions is very often awfully simplistic and terribly revealing: bad guys use illegitimate violence, then good guys use legitimate violence, and they win. After that, "spin doctors" have an all too easy job of molding into this pattern the diverse story-telling scenarios required for "just wars," for violent criminal justice methods, or for global surveillance. Our entertainment system is an immense tool for normalizing confrontation and justifying violence. It does not openly promote violence: for this reason our sensitivity to violence is neurotic, manipulated by *double-bind* messages (in the sense of the Palo Alto school of systemic psychology). Sapiential technoethics offers a possibility of cure, at least for one's personal interface with the infosphere, through the cultivation of a radical aversion to violence, even in representation – the representation of violence being an act of violence itself.

To take the most popular form of entertainment, TV series in their majority are murder stories or more precisely criminal police investigations – the "show" opens with a crime, most of the time sordid. How can people find it enjoyable to spend their evening on crime scenes, in a judiciary dissection facility, or in the company of mafiosi? A strange moral fragmentation is here at work: no bond is felt with the real sufferings caused by the corresponding real-world events. Reorienting a split mind toward aversion to violence can be powerful. It requires specific wisdom virtues, notably humility and courage: a lot of people do not even imagine that one can systematically avoid watching violence and that there is a message in doing so. Doing so and letting it be known could orient creative talents toward different inspirations. In the end, it can be hoped that the talents invested in a violent series like *Games of Throne* will be excellent as well in a story centered on non-violence and the benefits of *not* running for power, instead of exhibiting violence and the harm caused by obsessive running for power. The old TV series *Kung Fu* (starring David Carradine, non-Asian and inept in martial arts) apparently promoted an Eastern form of non-violence in the midst of paradigmatic "Western" violence (the Frontier). In fact, the episodes were framed all over again by the same scenario of justified violence (bad guys used illegitimate violence, then the good guy uses legitimate violence and he wins).

Aversion to violence, even in representation, would apply to news reports, which obviously count on its attractive power, and to video games. The question is not about *causing* violence in the real world. It is not about the mechanical or allegedly subliminal influence of violence representation on real world violence. It is more basic and more ethical. It is about one's relationship with violence in itself: educating sensibility and moral judgment in order to reshape a civilization of normalized and unseen violence. Violence in representation and the undercover promotion of violence are so pervasive in the current infosphere that it requires a real virtuosity for the modern sage to detect and evade it.

Ethical Skills: Non-confrontation, Non-violence, Frugality (Temperance)

Non-confrontation is an ethical ideal larger and deeper than non-violence. The purpose of introducing this notion is to modify the basic routines in our prehistoric "mental Operating System," so to say, instead of barely implementing non-violence as a new "mental software." I argue that confrontation is not the standard procedure to deal with the world in general and with other human persons in particular. Our prehistoric brain is apparently acting under the opposite instruction; it perceives confrontation and triggers confrontation far too easily.

The wisdom practices of harmony, humility, benevolence and courage compose non-confrontation as a specific moral disposition. Awareness is also needed to appreciate a situation as a systemic whole where no opposite sides need to be confronted. Autonomy allows us to be consistent without joining one side of a conflict, which means idly merging into its collective identity. Confrontation forges a group identity and for this reason the autonomy of the self must be a capacity to be consistent independently of confrontation patterns. Weak selves need opponents and allies for knowing and feeling who they are. In a lot of psychological and social circumstances they would initiate conflict, looking for identity through confrontation. Non-confrontation dismisses this form of moral dereliction.

The acceptability of confrontation may vary according to contexts and persons, but we are still in a world where confrontation can appear everywhere at any moment. In the infosphere in particular, radically divergent views, and usually hostile ones, are at one click away from each other. Conversation can be submerged in permanent aggression and counter-aggression on any forum – this digital aggression is called "flaming" (mail) or "trolling" (forum).

The Stoic doctrine of the *adiaphora* widens the common perception of good and evil (Cooper 2012: Sections 4.6 and 4.7): a lot of things are neither good nor evil, in a robust moral meaning of these terms. Instead, they are morally "indifferent" (*adiaphora* in Greek). They can be used for good when they are present but for evil also. When they are absent, the good life is still possible, at least for the sage. Health and wealth are major *adiaphora*; mobile phones and payment cards can be modern *adiaphora*. Probably most of the gadgets in the technosphere can be considered *adiaphora*, they are morally indifferent, despite their being morally significant. The law of Kranzberg applies here with its two parts (Section 1.1): "Technology is neither good nor bad; nor is it neutral." The ethical non-neutrality of artifacts means here that the self's attitudes are what matters. More precisely the self's meta-attitudes are what matters and what can make technology morally significant while being morally indifferent. A Zen position again.

Daoist non-action and its derivative tenets further explain the paradox of non-confrontation as transformative engagement. The seminal reference is Lao Zi's principle of victory without a fight (without confrontation): "It is the way of Heaven not to strive, and yet it skillfully overcomes" (Lao Zi 500 BCE: Section 73). The Dao prevails without fighting; the sage can and must do the same. This version of non-confrontation is recurrent in Chinese (both Daoists and Confucians) and Buddhist (mostly Zen) classics. Jullien (1998) contends that the sage has no idea, holds no position, and enters into no confrontation in Chinese classical thought. Jullien's explanation of this attitude is strikingly close in many passages to Heidegger's *Sein lassen* ("letting be") and serenity ("letting go").

In his adaptation of Buddhist doctrines to modernity, Harvey emphasizes the transformative power of non-confrontation. He quotes the Dhammapada: "Enmities never cease by enmity in this world; only by non-enmity do they cease" (Harvey 2000: 239). Harvey affirms that non-violence and non-confrontation consist in just not accepting the confrontation (Harvey

2000: 246). This formulation is valid for self-defense in aikido and similar techniques: confrontation is "proposed" by the other, but there is no necessity to "accept" it – contrary to the immediate reaction of our prehistoric brain. Aikido maintains that a proposition for confrontation can always be reinterpreted from the point of view of harmony: perceiving the global situation (oneself, the opponent, the environment) as a case of harmony, with internal potentials for evolution toward a preferable state through minimal action at the appropriate place and moment (Ueshiba 2007). In wise self-defense, the lowest level of resistance and opposition (ideally: pure non-action) is the best means to realize the desired state – which is not directly the defeat of the opponent but the end of the opposition, a return to a higher state of harmony. When receiving an aggressive email, the best means for returning to harmony on both sides is to send no answer; this non-action is preferable to any possible answer in many cases. The deplorable phenomenon of "flaming" (rapidly upgrading controversy) in electronic media is not an effect of the technology but of the first reaction of the prehistoric brain: accepting confrontation. I suggest systematically checking "nothing" as the first item on our lists of "What to do now in response?". Simple Zen.

Non-confrontation perhaps is active in the wise resistance to marketing pressure in the case of electronic tablets (like the iPad). Although they have been heavily promoted by the media since 2010, many independent observers think that they are largely a marketing manipulation, destined to compensate for the decreasing sales of personal computers, which itself is due to the saturation of the market by sufficiently performing devices. But satiety makes no money. The tablet promoted as the obligatory successor of the laptop computer is objectionable because it tries to pull the user away from the human and social flourishing possibilities of personal computers in the direction of media consumption and proprietary software dependency. The wise reaction is perhaps not an outburst against capitalist and corporate conspiracy but non-controversial patience, until the phenomenon dies by itself. Just do not feed the beast by contemplating buying one. Awareness of the significance of a microcomputer as a generative tool, courage not to follow the consuming hordes, and more generally an authentic appropriation of technology are all components of an ethics of constructive resistance and not of confrontation, in this precise case.

Defusing confrontation is an art, which the modern sage tries to master as an all-purpose existential virtuosity. Non-confrontation is a meta-attitude that leads to non-trivial appreciation and action in the technosphere and infosphere. This meta-attitude neutralizes the opposite meta-attitude of confrontation that characterizes modern technology in its entirety.

Non-violence is naturally a virtue and a skill highly praised and staunchly applied by the modern sage. This practice depends on the solid background of non-confrontation, which remains a broader and deeper wisdom virtue. It is also dependent on the specific sensitivity to any form of violence that has been described as a requisite for wisdom practices. In this context, non-violence for the modern sage consists neither in wishful thinking – the moral condemnation of violence – nor in heroic cases – where non-violence is an "extraordinary" brave attitude in face of injustice.

Buddhism begins with the awareness of suffering. One of the best-known consequences of this awareness is the practice of non-violence. The master of non-violent action in modernity is of course Gandhi, who was not a Buddhist but a Hindu, but whose inspiration was in the broad spiritual traditions of Indian thought as a whole. Gandhi's non-violent ethos burst into Western politics as an episode of dewesternization. Naess extends Gandhi's methods of protest and action on public policies in numerous books and articles, most of them published in volume 5 of his *Works* (Naess 1974). With Naess, the scope of non-violent action becomes the entire field of civilizational issues in the contemporary. The philosophical robustness of non-violent action is also reinforced in Naess by an original theory of self-realization inside the harmonies of nature. More than a renovation of political militancy, Naess's philosophy calls for a self-practice of wisdom.

However, issues of non-violence are commonly construed as political questions, concerning foreign or state politics and policy. It is also in this institutional context that violence between persons is regarded as criminality, an issue of law enforcement. A renewed ethic, focusing on self-constitution, will consider violence and non-violence as questions of care – care for oneself, care for other selves and persons. In real life, non-violence is almost never heroic behavior in an extraordinary case but rather everyday care for human persons, including one's self.

Moral violence matters. The question of reflexive and self-inflicted moral violence must not be underestimated. The philosophical cause of this problem is the heritage of the culture of guilt from religious and moral ideologies. The pragmatic effect of this phenomenon is ordinary suffering: at work, at school, and in the family circle. On this basis, selves who are harassed by moral violence are prone to exert moral or physical violence toward others. It originates in self-inflicted moral violence: guilt, self deprecation, and personal dereliction directly due to moral domination episodes.

Physical violence comes next in the order of contemporary problems. The new awareness that technoethical wisdom tries to create should reveal the systematic violence of our technology. Following Table 1.3 ("The Three Kinds of Power"), technology is the first form of power: domination over things. Weapons and wars are linked also to the second form of power in Table 1.3: domination over humans. Be it over nature or over humans, open violence dominates our relationship with the world. Most of the time it is implicit and latent violence and for this reason it remains unseen.

My demonstrative instance will be the implicit violence in the single object that founds the constitutive cult of Western civilization: the automobile. Road traffic violence is a typical case of violence in the technosphere, not particularly in "road rages" (people fighting each other after a traffic incident) but in ordinary driving and in the simple figures of road casualties. Driving has been mentioned above as a possible field to exercise benevolence. The latent aggressive mood in a lot of drivers seems to be specifically triggered by the fact of driving. In a technoethical perspective, the car appears as the perfect dominated and dominating object. A phenomenology of driving would consider it the perfect device for dominating the world: space and time, one's physical environment, and the other humans, symbolically. Beyond this personal psychological effect, the acceptability of road traffic casualties is a profound mystery; the "car cult" hypothesis sadly explains it (cults can be bloodthirsty). Traffic kills and wounds more than firearms; everyone in an industrial country knows someone who has been killed or severely injured on the road. This ordinary violence is rarely taken for what it is.

How pervasive and systemic the culture of violence is can be illustrated by the first few minutes of the US fictional series, *House of Cards*. Wikipedia gives the following account:

> The series opens with congressman Francis Underwood displaying his ruthless practicality by killing a suffering pet dog with his bare hands while explaining to the audience how there are times when we require someone to do the unpleasant thing yet the necessary thing.
>
> (Wikipedia: "House of Cards")

All the violence is "out of the frame," unseen but typical. First, road violence hurting the dog, and then the "necessary" or justified violence in return: killing the dog (with bare hands, guns are not violent enough). This is exactly the standard scenario that I pinpointed above when discussing non-confrontation. Is the killing "unpleasant" as the brave character pretends? There is a feeling that it is profoundly rewarding to be put in a situation where violence is justified. As for Francis Underwood, I cannot tell because my watching the series definitely ended just after this scene.

Frugality is the practice of wisdom that will produce the most visible effects in the technosphere. It also demands the most urgent change in lifestyle considering the current situation of (relative) abundance and massive consumption – for the 80 percent in industrialized countries at least. Frugality means neither abstinence nor poverty. It means the mastery of one's level of satisfaction, first by a precise and authentic awareness of what satisfaction is and then by a capacity to limit one's intake according to one's best judgment. The essential in frugality is not that the person's level of satisfaction is low. It is recognizing that there is a level of satisfaction, that the person is aware and responsible for it, limiting her intake on abundance to that level. It applies to what people can buy in a supermarket with the amount of money they have, to what they can eat at any moment from the stock in the kitchen cupboards and fridge, to the house or the car they can buy by stretching their borrowing capacity to its maximum. A high level of consumption can be either assumed as a "necessity" or reconsidered: taking into account all the consequences of this demanding lifestyle, the obligation to earn the money for it and at what cost for one's life. Frugality only means that there is a limit.

Sobriety is a fundamental precept in Buddhist ethics from the beginning (Harvey 2000: 77). Zen radicalized this tendency sometimes in a quasi-mystical quest. A Zen esthetics of frugality animates Thoreau's *Walden*. Simplicity is the first virtue mentioned in *Walden* (Thoreau [1854] 1985: 334). Thoreau's broader argument relies on one of the astonishing affordances of modernity: being a primitive in it, which is not at all the same as leaving modernity, refusing it, or fighting against it. Instead, being primitive in modernity means dwelling in it with specific existential skills. Frugality presides over this existential virtuosity.

I have often mentioned Gandhian economics and Gandhian simplicity as inspirations for technoethics. Both plead for a frugality that does not shun modernity or even (relative) abundance. A famous maxim attributed to Gandhi is that there is enough to satisfy every man's need but not every man's greed. It calls for frugality, the contrary of greed, but not for deprivation, the contrary of satisfying one's needs. From a rather abstract but illuminating point of view, Vasanthi Srinivasan assimilates Gandhi's *humble practices* with Borgmann *focal practices* (Srinivasan in Hershock et al. 2003: 310). She intends to show that Gandhian frugality is an indirect response to Borgmann's quest for authenticity in the midst of the technosphere. The primary goal of her article is even more ambitious. She intends to show that Gandhian frugality can be the "saving power" that Heidegger was expecting as a response to modern technology. I tend to agree.

For implementing this power, the concept of *satiety* (having enough) is central to frugality as existential virtuosity. The modern sage tracks the feeling of satiety in the most ordinary experiences of life in order to cultivate it and to become ever more aware and responsible for it. Food is the easiest domain to begin practicing frugality before expanding the skills for satiety to all the other dimensions of existence, including money and vanity. The capacity for satiety and frugality genuinely "disencumbers" the access to the good life in the technosphere. Ironically, abundance hinders flourishing; facilitation makes things harder for self-construction. The everlasting desire to have more makes enduring satisfaction impossible. The capacity to have enough is a relief

The difference between *use* and *abuse* of addictive substances and practices gives an idea of the spirit suggested by modern wisdom: aware, careful, but constructively open to existential experiences and possibly collective experiences of sharing. The culture of alcohol drinking in some modern societies and the culture of other addictive substances in other societies entail educating satiety. It requires a self-culture of frugality, as a culture of the self, and not only the right and safe use of an addictive substance or practice. This case demonstrates that frugality does not mean *abstinence.* Frugality is the capacity to drink alcohol without being an alcoholic, not the capacity never to drink alcohol. Virtue is a middle term. It depends on a wise judgment adapted to the circumstances, and it is the rule for practical judgment, Aristotelian *phronēsis*, *Zhongyong*'s middle way, and wisdom's existential virtuosity.

Another instance is shopping, ordinary shopping: the weekly supply of food, small commodities, possibly clothes. It should be possible to morally endorse its pleasures, or at least the pleasure of reassurance that shopping brings in ordinary life. In a subtle balance on the edge of addiction, shopping can be experienced as neither a recurrent chore of modern life nor the innocent delight of abundance. It can be something in the middle and very different in nature: an existential exercise in the whole set of wisdom practices and in particular in the research of one's personal version of frugality.

How many shoes do we need? Most humans have two feet and only two. The different circumstances that impose specific shoes, for material or social reasons, could hardly reach the dozen. In owning more than a certain number of shoes something different is at stake, something concerning the self and its identity. This personal concern (the self and its identity) can be addressed through shoes possession but this kind of issue is better dealt with on its specific level (psychological, moral, social) instead of being dealt with in terms of shoes or clothes. Improving one's consistence and self-reliance should lead to a lesser dependency on "status commodities" and the skills in this domain constitute frugality.

Useful Practices: Meditation, Life Hygiene, Bodily Maintenance

Ordinary self-care is the final step in this analysis of modern wisdom and yet it may be the first step in personal reform, for living differently in the technosphere.

Meditation is essential self-care. In a negative and quasi-definitional sense, meditation is simply a moment when we do not interact with anything, not even language. This Section concentrates on "pure" meditation: meditation as an exclusive activity during a period of time (different from *samu*, meditation within an activity). In Asian doctrines of wisdom, this form of meditation is a necessary practice; it is even the almost exclusive practice in some Zen schools, such as the Soto school and its *shikantaza* orientation – *shikantaza* means "simply sitting." It defines meditation as the most simple possible activity or inactivity: simply sitting and breathing.

Taking a moment in one's daytime (or nighttime) and simply sitting, lying, or walking (in some Zen schools) while clearing out one's mind of every thought, non-confrontationally, by letting any mental representation just go its way and leave: this is not a complex technique, neither to learn nor to practice. If one discovers that one does not have a moment in life to take such a break, then awareness of the "time war" and of personal alienation may surface. When practicing meditation, progress is usually smooth and rewarding, even if nothing spectacular happens.

Remaining focused on the ordinary is a key principle both for tech-noethics and for meditation. Sitting and breathing are extremely ordinary activities. For this reason, identifying them as values and mindfully practic-ing them is an existential virtuosity. Numerous websites display informa-tion about ancestral and modern meditation techniques but only the best ones insist enough on the capital idea that nothing technical is important for meditation – neither position nor breathing technique, not even mental techniques. Meditation then appears as the attitude *par excellence* to free oneself from anything technical and technological, in the midst of the technosphere. Mastering this detachment is a resource for mastering attachment.

When Hershock (1999) gives an account of the original doctrines of Zen in opposition to the modern system of values, the main point is about control, the obsession of modernity. The evolution suggested by Hershock is from control to active meditation; he often formulates it as a change from control to appreciation. Control is for him a real addiction, similar to drug addiction (Hershock 1999: 151–154), even if its vector is the "soft" power of the media (Hershock 1999: Chapter 6). The impoverishment of life in the technosphere is described through ancient Buddhist narratives about human dereliction, composing a vivid picture of how we are "losing touch with what matters" (Hershock 1999: 47):

> Our relationship with the things around us has ceased to be a conver-sation – a turning together in creative reciprocity – and degenerated into the worst kind of monologue: the giving of orders.
>
> (Hershock 1999: 198)

Ordinary meditation can treat addiction and restore wisdom in the techno-sphere. It can consist in doing nothing but doing it consciously and intensely. "A standard Chan definition of meditation was *meeting situ-ations without obstruction*," recalls Hershock (2005: 143). This is a supreme mastery of non-confrontation and a topical wisdom exercise in the context of modernity: the exact opposite of control.

Life hygiene, including diet and the management of addictive sub-stances, was the essential part of preventive medicine in Antiquity. We are discovering again the obvious advantages of preventive medicine over cur-ative medicine. They are all the more obvious for the diseases currently deprived of curative treatments. Preventive medicine and health care in general belong to modern wisdom. The technosphere is full of resources for the motivated self in matters of life hygiene and ordinary self-care. The abundance of easily retrievable information (from the Web) and the (rel-ative) abundance of goods – drinkable water to start with – and facilities – sanitation in particular – give us the unique opportunity to live in a technosphere which is more fit than ever for a healthy life.

The key instance of sapiential life hygiene seems to be, as it has always been, "conscious" eating, but now in a renewed, deepened, and broadened meaning that includes being aware of and caring for an extensive set of harmonies in one's body, in the local and global environment, and also in the economy, local and global.

As it bears on daily trivial details, life hygiene remained for a long time a subject about which the average human in industrialized countries did not meditate on or even think much about. Ordinary behavior was driven from the outside, by the advertisement industry and by government directives in the end. In the industrial period, governmentality imposed this sort of biopower on people, as we know from Foucault. Modern domestic behaviors have been voluntarily implanted. Now is the time for self-reliant norms of life hygiene, which implies a reappropriation of biopower, a bio-empowerment of the self. The infosphere is of irreplaceable help for all those engaging a personal reform in life hygiene.

Bodily maintenance by exercise, from daily gym to martial arts, without any "sport" meant as competition, can be taken as a humble but exacting existential virtuosity. It belongs to life hygiene but exceeds its utilitarian goals. For personal reform and practical wisdom, the tragic dichotomy between mind and body in Western culture must simply be abandoned. Unity and harmony between mind and body was a pillar of ancient cultures. Modernity, in the industrial era, has been built on dualism, the separation of mind and body, justifying the exclusive ethical importance of the mind, which was supposed to be the only immortal or divine part of the human.

Modern wisdom requires that we discover again the source of joy and pleasure, easily shared with others, that physical activity is and should always have been. Technology has no direct power to restrain us from physical activity. On the contrary, it provides countless means and incentives for physical exercise, if only we manage our interface with the technosphere according to technosapiential principles. To argue the first point (technology implies no restriction in physical activity), I favor one more time the case study of the car, our cult object: it is sometimes said to "prevent" us from walking. This is literally not true: chains and shackles would do that, a car cannot. A close inspection of every part and function in a car will find nothing destined to impede the walking of a human being. The fact is that we have always-already (as existential analysis characteristically says) chosen driving instead of walking. Since we are not in a reflexive and sapiential attitude we transfer this (moral) decision to the car itself – or with even more hypocrisy graduates would transfer it to the "system." The same analysis applies to the elevator/stairs case study and to many similar cases. To argue the second point (the technosphere is full of incentives for physical activity), I will mention affordable running shoes, bicycles, clothes, and all kinds of sport equipments and devices, and even gadgets such as the smartphone apps for monitoring one's jogging, or paced breathing for relaxation or sport, and many others. A complex technosystem like an exercise bicycle equipped with a video device (a laptop computer will do) provides basic healthcare and personal satisfaction at the same time, mind and body indistinctly.

Beyond these easy opportunities of bodily maintenance (mind-body maintenance actually), more demanding physical activities can be invested in a self-construction project. Martial arts are evidently the privileged practice for a sapiential quest. Their presence all over the world, in different modern cultures, proves that mind-body education is still possible and can be understood as a flourishing opportunity for all. But five minutes of gym exercises twice a day is enough to remind the post-industrialized human being that one has a body and that one is a body (*tai chi* or *qi gong* techniques for this five minutes will help). The well-being sensation in one's shoulders, neck, or waist muscles is a significant, discreet, and lasting reminder of one's existential consistence. These practices and feelings are existential virtuosity; they enrich intimate life and help give it a decisive turn toward practical wisdom. The present technosphere and infosphere does not make them more difficult but easier.

A Final Word on Sapiential Technoethics

The ambition of this book was to give one version of a philosophy and an ethics of contemporary technology. This project leads to a stance that can be characterized by the following properties:

1 *Practical wisdom* is definitely recommended for overcoming both the deficit of action in the intellectual world and the deficit of intellectual ambition in pragmatic endeavors. The gap between theory and practice has never been so tragic: theory is over-sophisticated and the privilege of an elite; practice is over-powerful and in the hands of everyone.
2 A *specific* ethics is offered for the technosphere and the infosphere. The history of philosophy provides the means for it but it did not deliver a specific set of values applying to a post-industrial and digital environment.
3 The *focus on the ordinary* gives a new impulsion to technoethics. The empowerment of individuals in the present technosphere and infosphere bestows agency on ordinary microactions, a level that remains underestimated in established humanities and social sciences.
4 *Diversity* in technoethics is in tune with the globalized technosphere. On the one hand, non-Western approaches to reality and to wisdom contain the inspirations for a new balance, and, on the other hand, individual "customizations" of wisdom, inventive and collaborative, contain the generativity for a better technosphere, largely through a better infosphere.
5 *Ambition and humility* go together for surviving in an intellectual landscape that has been left devastated by the experts of critique. At first sight, pleading for wisdom in the contemporary world may seem "vain": useless and arrogant at the same time. This book has attempted to arouse second thoughts about established pessimism in defending constructive technoethics as a humble ambition.

References

Ames, Roger T., and David L. Hall. 2001. *Focusing the familiar: A translation and philosophical interpretation of the Zhongyong*. Honolulu: Hawaii University Press.

Angle, Stephen C. 2009. *Sagehood: The contemporary significance of Neo-Confucian philosophy*. Oxford: Oxford University Press.

Angle, Stephen, and Michael Slote. 2013. *Virtue ethics and Confucianism*. New York: Routledge.

Aubenque, Pierre. 1963. *La prudence chez Aristote*. Paris: PUF.

Barry, John. 2006. Resistance is fertile: From environmental to sustainability citizenship. In Dobson, Andrew, and Derek Bell (eds), *Environmental citizenship*, pp. 21–48. Cambridge, Mass.: MIT Press.

Bartsch, Shadi, and David Wray (eds). 2009. *Seneca and the self*. Cambridge: Cambridge University Press.

Benkler, Yochai. 2011. *The Penguin and the Leviathan: How cooperation triumphs over self-interest*. New York: Random House.

Borgmann, Albert. 1984. *Technology and the character of contemporary life: A philosophical inquiry*. Chicago, Ill.: University of Chicago Press.

Borgmann, Albert. 2006. *Real American ethics: Taking responsibility for our country*. Chicago, Ill.: University of Chicago Press.

Boyd, Danah Michele. 2014. *It's complicated: The social lives of networked teens*. New Haven, Conn.: Yale University Press. www.danah.org/books/ItsComplicated.pdf.

Briggle, Adam, and Carl Mitcham. 2009. Embedding and networking: Conceptualizing experience in a technosociety. *Technology in society* 31(4): 374–383. http://dx.doi.org/10.1016/j.techsoc.2009.10.001.

Bynum, Terrell Ward. 2006. Flourishing ethics. *Ethics and information technology* 8(4): 157–173.

Cafaro, Philip. 2004. *Thoreau's living ethics: Walden and the pursuit of virtue*. Athens, Ga.: University of Georgia Press.

Chan, Joseph. 2014. *Confucian perfectionism: A political philosophy for modern times*. Princeton, N.J.: Princeton University Press.

Cooper, John M. 2012. *Pursuits of wisdom: Six ways of life in ancient philosophy from Socrates to Plotinus*. Princeton, N.J.: Princeton University Press.

Darling-Smith, Barbara. 2002. *Courage*. Notre Dame, Ind.: Notre Dame University Press.

Dewey, John. 1920. *Reconstruction in philosophy*. New York: Henry Holt.

Dewey, John. 1984. *Individualism, old and new* [1930]. In *The later works*, Carbondale, Ill.: Southern Illinois University Press, vol. 5, pp. 41–123.

Dewey, John, and James Hayden Tufts. 1908. *Ethics*. New York: Henry Holt.

Emerson Ralph Waldo. 1870. *Society and solitude*. Boston, Mass.: Houghton Mifflin. Repr. New York: Cosimo, 2005. www.rwe.org/complete-works/vii-society-and-solitude.html.

Emerson, Ralph Waldo. 1983. *Essays and lectures*, ed. J. Porte. New York: The Library of America.

Ess, Charles. 2009. *Digital media ethics*. Cambridge: Polity Press.

Foucault, Michel. 2001. *L'herméneutique du sujet. Cours au Collège de France (1981–1982)*. Paris: Gallimard/Seuil. (*The hermeneutics of the subject: Lectures at the Collège de France, 1981–1982*. New York: Picador, 2006).

Frankfurt, Harry G. 2006. *Taking ourselves seriously: Getting it right.* Redwood City, Calif.: Stanford University Press.

Geerts, Robert-Jan. 2012. Self practices and the experiential gap: An analysis of moral behavior around electricity consumption. *Techné* 16(2): 94–104.

Gordhamer, Soren. 2008. *Wisdom 2.0: Ancient secrets for the creative and constantly connected.* New York: HarperOne.

Grinevald, Jacques. 1976. La révolution carnotienne. Thermodynamique, économie et idéologie. *Revue européenne de sciences sociales, Cahiers Vilfredo Pareto* 36: 39–79.

Harvey, Peter. 2000. *An introduction to Buddhist ethics: Foundations, values and issues.* Cambridge: Cambridge University Press.

Heidegger, Martin. 1939. Vom Wesen und Begriff der Physis. Aristoteles, Physik B, 1. In *Wegmarken*, Frankfurt a.M.: Klostermann, 1967, 237–299. (*Pathmarks.* Trans. W. McNeil. Cambridge: Cambridge University Press, 1998).

Heidegger, Martin. 1959. *Gelassenheit.* Pfullingen: Neske. (*Discourse on thinking.* Trans. J.M. Anderson. London: Harper & Row, 1969).

Herrigel, Eugen. 1953. *Zen in the art of archery.* Trans. R.F.C. Hull. Pantheon. Repr. Vintage Spiritual Classics, 1989.

Hershock, Peter D. 1999. *Reinventing the wheel: A Buddhist response to the information age.* Albany, N.Y.: State University of New York Press.

Hershock, Peter D. 2005. *Chan Buddhism.* Honolulu: Hawaii University Press.

Hershock, Peter D. 2006. *Buddhism in the public sphere: Reorienting global interdependence.* London; New York: Routledge.

Hershock, Peter D. 2012. *Valuing diversity: Buddhist reflection on realizing a more equitable global future.* Albany, N.Y.: State University of New York Press.

Hershock, Peter D. 2014. *Public Zen, personal Zen: A Buddhist introduction.* Lanham, Md.: Rowman & Littlefield.

Hershock, P.D., M. Stepaniants, and R.T. Ames (eds). 2003. *Technology and cultural values: On the edge of the third millennium.* Honolulu: Hawaii University Press.

Hursthouse, Rosalind. 1999. *On virtue ethics.* Oxford: Oxford University Press.

Irvine, Ben. 2011. Wisdom and the good life: A philosophy of conscience. *Journal of modern wisdom* 1: 4–11.

Ives, Christopher. 1992. *Zen awakening and society.* Honolulu: University of Hawaii Press.

Jullien, François. 1997. *Traité de l'efficacité.* Paris: Grasset. (*A treatise on efficacy: Between Western and Chinese thinking.* Trans. Janet Lloyd. Honolulu: University of Hawaii Press, 2004).

Jullien, François. 1998. *Un sage est sans idée, ou l'autre de la philosophie.* Paris: Seuil.

Jullien, François. 2005. *Nourrir sa vie. A l'écart du bonheur.* Paris: Seuil. (*Vital nourishment: Departing from happiness.* Trans. Arthur Goldhammer. Brooklyn: Zone Books, 2007).

Kane, Robert. 2010. *Ethics and the quest for wisdom.* Cambridge: Cambridge University Press.

Kekes, John. 1995. *Moral wisdom and good lives.* Ithaca, N.Y.: Cornell University Press.

Knappenberger, Brian. 2014. *The Internet's own boy: The story of Aaron Swartz* (documentary film, 105 mins). Luminant Media. https://archive.org/details/TheInternetsOwnBoyTheStoryOfAaronSwartz.

Kohlberg, Lawrence. 1981. *The philosophy of moral development: Moral stages and the idea of justice.* San Francisco, Cal.: Harper & Row.

Korsgaard, Christine M. 2008. *The constitution of agency: Essays on practical reason and moral psychology.* Oxford: Oxford University Press.

Lao Zi. 500 BCE. *Dao de jing,* Chinese and English. http://ctext.org/dao-de-jing.

Lao Zi. 1996. *Lao Tzu's Tao-Teh-Ching: A parallel translation collection.* Edited by B. Boisen. Boston, Mass.: GNOMAD Publishing.

Margalit, Avishaï. 1996. *The decent society.* Cambridge, Mass.: Harvard University Press.

Maxwell, Nicholas. 2007. *From knowledge to wisdom: A revolution for science and the humanities.* Oxford: Blackwell, 1984. 2nd edn, London: Pentire Press.

May, Reinhard. 1989. *Ex Oriente Lux. Heideggers Werk unter ostasiatischem Einfluss.* Stuttgart: Steiner. (*Heidegger's hidden sources: East Asian influences on his work.* Trans. G. Parkes. London: Routledge, 1996.)

Musō, Kokuchi (Soseki). 1994. *Muchū Mondō [13–14th century]/Dream conversations: On Buddhism and Zen.* Trans. T. Cleary. Boston and London: Shambhala.

Naess, Arne. 1974. *Selected works, vol V: Gandhi and group conflict. Explorations of nonviolent resistance, Satyagraha.* Revised edn, 2005. Dordrecht: Springer.

Naess, Arne. 2005. *Selected works, vol. X: Deep ecology of wisdom: Explorations in unities of nature and cultures.* Dordrecht: Springer.

Nhat Hahn, Thich. 2000. *Essential writings,* ed. Robert Ellsberg. Maryknoll, N.Y.: Orbis Books.

Nhat Hahn, Thich. 2005. *Being peace* (new edition). Berkeley: Parallax Press.

Nozick, Robert. 1989. *The examined life: Philosophical meditations.* New York: Simon & Schuster.

Okakura, Kakuzō. 1906. *The book of tea.* Dreamsmyth edition, 2011. www.gutenberg.org/ebooks/769.

Parkins, Wendy, and Geoffrey Craig. 2006. *Slow living.* Oxford; New York: Berg.

Pirsig, Robert M. 1974. *Zen and the art of motorcycle maintenance.* New York: Bantam.

Poceski, Mario. 2007. *Ordinary mind as the way: The Hongzhou school and the growth of Chan Buddhism.* Oxford: Oxford University Press.

Pye, Michael. 2003. *Skilful means: A concept in Mahayana Buddhism.* 2nd edn. London: Gerald Duckworth.

Rahula, Walpola. 1959. *What the Buddha taught.* New York: Grove Press.

Reeve, Charles D.C. 1992. *Practices of reason: Aristotle's "Nicomachean ethics."* Oxford: Clarendon Press.

Reeve, Charles D.C. 2013. *Aristotle on practical wisdom: Nicomachean Ethics VI.* Cambridge, Mass.: Harvard University Press.

Rifkin, Jeremy. 1987. *Time wars: The primary conflict in human history.* New York: Simon and Schuster.

Rifkin, Jeremy. 2009. *The empathic civilization: The race to global consciousness in a world in crisis.* Cambridge: Polity Press.

Robinson, David M. 1993. *Emerson and the conduct of life: Pragmatism and ethical purpose in the later work.* Cambridge: Cambridge University Press.

Rushkoff, Douglas. 2010. *Program or be programmed: Ten commands for a digital age.* New York: OR Books.

Schürmann, Reiner. 1987. *Heidegger on being and acting: From principles to anarchy.* Bloomington, Ind.: Indiana University Press.

Seneca. 64 AD. *Moral letters to Lucilius*. Trans. R.M. Gummere. Loeb Classical Library. 1917–1925. http://en.wikisource.org/wiki/Moral_letters_to_Lucilius.

Shantideva. 763 AD. *Bodhicaryâvatâra*. www.berzinarchives.com/web/x/pdf/?type =pdf&book=true&path=/web/x/prn/p.html_1487505749.html&__locale=en.

Slingerland, Edward. 2003. *Effortless action: Wu-Wei as conceptual metaphor and spiritual ideal in early China*. Oxford: Oxford University Press.

Slingerland, Edward. 2015. *Trying not to try: Ancient China, modern science, and the power of spontaneity*. New York: Broadway Books.

Spence, Edward H. 2011a. Information, knowledge and wisdom: Groundwork for the normative evaluation of digital information and its relation to the good life. *Ethics and information technology* 13(3): 271–275.

Spence, Edward H. 2011b. Is technology good for us? A eudaimonic meta-model for evaluating the contributive capability of technologies for a good life. *Nanoethics* 5: 335–343.

Sternberg, Robert J. (ed.). 1990. *Wisdom: Its nature, origins and development*. Cambridge: Cambridge University Press.

Stoehr, Taylor. 1979. *Nay-saying in Concord: Emerson, Alcott, and Thoreau*. Hamden, Conn.: Archon Books.

Tavani, Herman T. 2004. *Ethics and technology: Ethical issues in an age of information and communication technology*. 3d edn. Hoboken, N.J.: John Wiley & Sons.

Thoreau, Henry David. 1985. *A week on the Concord and Merrimack rivers [1849]; Walden or life in the woods [1854]; The Maine woods; Cape Cod*. New York: The Library of America.

Thoreau, Henry David. 2004. *The higher law: Thoreau on civil disobedience and reform*, ed. W. Glick. (H.D. Thoreau, *The writings*), Princeton, N.J.: Princeton University Press.

Ueshiba, Morihei. 2007. *The art of peace*. Trans. J. Stevens. Boston; London: Shambhala.

Van Cromphout, Gustaaf. 1999. *Emerson's ethics*. Columbia, Mo.: University of Missouri Press.

Verbeek, Peter-Paul. 2013. Resistance is futile: Toward a non-modern democratization of technology. *Techné* 17(1): 72–92. http://dx.doi.org/10.5840/techne 20131715.

Wei-Hsun Fu, Charles, and Sandra A. Wawrytko (eds). 1991. *Buddhist ethics and modern society*. New York: Greenwood Press.

Wong, Pak-Hang. 2011. Dao, harmony and personhood: Towards a Confucian ethics of technology. *Philosophy and technology* 2011. www.springerlink.com/content/u667060p46746115.

Yamamoto, Jōchō (Tsunetomo). 1979. *Hagakure: The book of the samurai*. Trans. W.S. Wilson. Tokyo: Kodansha International. www.eisshinryu.com/b-hagakure-chapter-1.htm.

Zhuang Zi. 475 BCE. *Zhuangzi*. Trans. James Legge. http://ctext.org/zhuangzi.

Zhuang Zi. 1968. *The complete works of Chuang Tzu*. Trans. B. Watson. New York: Columbia University Press.

Zhongyong. 600 BCE. *Zhongyong: The state of equilibrium and harmony*. Chinese and English. http://ctext.org/liji/zhong-yong.

Index